T0180415

Energy Conversion and Management

Giovanni Petrecca

Energy Conversion and Management

Principles and Applications

 Springer

Giovanni Petrecca
Department of Industrial Engineering
University of Pavia
Pavia, Italy

ISBN 978-3-319-34944-2   ISBN 978-3-319-06560-1 (eBook)
DOI 10.1007/978-3-319-06560-1
Springer Cham Heidelberg New York Dordrecht London

Printed on acid-free paper

Springer is part of Springer Science+Business Media (www.springer.com)

To my mother Valentina de Majo

Preface

After many years of researching, teaching, and consulting I am firmly convinced that all problems, in order to be solved, must first be reduced to their essentials.

That does not mean ignoring or skating over some aspects, but on the contrary it means going deeply into the core of the problem to understand it completely, and then focusing on the few elements really necessary for its solution.

This is quite a difficult approach because it requires a clear comprehension of what is really important and of what is not. Engineering universities generally train students to achieve that, but during the professional career a great effort must always be made to apply these principles to practical problems.

I became engaged in the research for a unifying approach to energy management after experiences in teaching electrical machines and drives with the unified theory, which in 1976 was rather innovative for the student curriculum, but was afterwards demonstrated to be an excellent approach to an intimate and all-embracing comprehension of the related phenomena.

Professional experiences in the field of industrial energy conversion and management had stimulated me to organize a course on this subject in the Engineering Faculty of the University of Pavia in Italy and, in 1992, to write a text providing a global view of energy management in industry and in buildings.

The stream of energy, in different forms, which is transformed step by step inside any factory and building, has been considered as the unifying factor for understanding all the topics, which at first glance seem to be comprehensible only in terms of independent theories.

After 20 years I have decided to write a new book that still keeps the original approach, but which also includes the innovations that have influenced this period.

In conclusion, this new book is not just another specialized textbook or handbook and so it does not compete with those already existing which remain indispensable tools for solving technical problems in detail. Rather, it aims to help the reader by giving them the fundamental elements to simplify and tackle problems within a global vision of the production units by exploiting the correlations always present among many topics apparently distant from one another.

Having completed this work, I should like to thank university and professional colleagues who made suggestions to facilitate the comprehension of many topics. In addition, thanks are due to CSE Srl, Pavia, Italy (founded in 1983, it now belongs to EDISON Group, Italy), from whose data bank most of the examples have been drawn.

Many thanks to all the students whom I have encountered over the years, as I have come to appreciate that teaching is the best way to deeply understand the subject you are talking about.

On a more personal note, I should like to thank my wife Marina, who helped me see the book through non-technical eyes.

Pavia, Italy   Giovanni Petrecca
February 2014

Contents

List of Figures

List of Tables

List of Examples

List of Symbols

Letter symbols include symbols for physical quantities and symbols for units in which these quantities are measured.

For symbol units see Tables 2.4, and 2.5.

Quantity symbols are listed below.

Efforts have been made to attribute a single-letter symbol to each physical quantity. Because of the number of topics introduced, some symbols have different meanings according to the context and to the common practice; in these cases, the list of symbols indicates the chapter to which these symbols apply.

A, S	Area, section
A_n	Transformer rated power
AE	Available energy (isentropic expansion)
AHU	Air handling unit
ASF	Actual steam flow
ASR	Actual steam rate
c	Specific heat
c_d	Dispersion coefficient (Chap. 13)
C	Capacitor (Chap. 7)
C	Heat capacity flow rate (Chap. 15)
C_p	Ratio between shaft power and power available in the wind (Chap. 4)
C_p	Present monetary value (Chap. 20)
COD	Liquid chemical oxygen demand
COP	Coefficient of performance (Chap. 12)
CU	Coefficient of utilization (Chap. 14)
d	Length (thickness, diameter)
E	Energy
E,P	Heat exchanger effectiveness (Chap. 13)
E_p	Electric active energy
E_q	Electric reactive energy
f	Electric frequency
f, f^*	Inflation rate (Chap. 20)
FT	Correction factor (Chap. 15)
FWF	Future worth factor
g	Gravitational acceleration

(continued)

(continued)

g_c	Conversion factor (Chap. 10)
h	Enthalpy per unit mass
h	Heat transfer coefficient (Chap. 8)
H	Head (Chap. 10)
H	Enthalpy
HHV	Higher (or gross) heating value
i	Coefficient for life cycle costing (Chap. 20)
I	Investment (Chap. 20)
I	Electric current
I_p	Present investment (Chap. 20)
IRR	Internal rate of return
ln	Natural logarithm
k	Thermal conductivity (Chap. 8)
k	Ratio of the gas specific heat parameter c_p/c_v at constant pressure (c_p) and volume (c_v) (Chap. 11)
K_c	Unburned combustible losses coefficient (Chap. 6)
K_m	Karman coefficient (Chap. 2)
K_s	Hassenstein coefficient (Chap. 6)
K_v	Velocity coefficient (Chap. 2)
l, L	Length
LHV	Lower (or net) heating value
m	Mass flow rate
M	Mass
n	Number of moles (Chap. 2, Table 2.5)
n	Life of investment (Chap. 20)
n	Number of phase conductors (Chap. 7)
n_v	Rate of ventilation (Chap. 13)
NACF	Net annual cash flow
NTU	Number of transfer units (Chap. 15)
ORC	Organic Rankine cycle (Chap. 9)
p	Pressure
ppm	Gas concentration (Chap. 6)
P	Power (mechanical, electric active power)
P, E	Heat exchanger effectiveness (Chap. 15)
PAF	Present annuity factor
P_{cc}	Transformer load losses (Chap. 5)
P_{cn}	Transformer load losses at rated power (Chap. 5)
P_o	Transformer no-load losses (Chap. 5)
PL	Electric line losses (Chap. 7)
PW	Present worth (Chap. 20)
PWF	Present worth factor (Chap. 20)
q	Volume flow rate
q_s	Volume flow rate in standard conditions
Q	Power as heat transfer rate

(continued)

(continued)

Q	Electric reactive power (Chap. 7)
Q	Annual revenues (Chap. 20)
r	Radius
r	Interest or discount rate (Chap. 20)
r	P_{cn}/P_o (Chap. 5)
r_p	Compression ratio (Chap. 11)
R	Electric resistance (Chap. 7)
R	Radius of gyration (Chap. 4)
R	Parameter for heat exchanger calculation (Chap. 15)
R	Universal gas constant (Chap. 2, Table 2.5)
RDF	Refusal-derived fuel
RH, ϕ	Relative humidity
ROR	Investors rate of return (Chaps 20 and 21)
$R_{th} = 1/U$	Overall thermal resistance (Chap. 8)
S, A	Section, area
t, T	Temperature (capital letter is generally used for absolute temperatures)
TOE	Ton oil equivalent
TSR	Theoretical steam flow rate (Chap. 9)
u	Internal energy (Chap. 11)
$U = 1/R_{th}$	Overall heat transfer coefficient (Chap. 8)
v	Specific volume
V	Voltage (Chap. 7)
V	Linear velocity
V	Volume (Chap. 2, Table 2.5; in Chap. 6)
VB	Volume of the building (Chap. 13)
x	Load factor (Chap. 5)
x	Steam quality index (Chaps. 6 and 9)
X	Electric reactance
Z	Compressibility factor
α	Instrument flow coefficient (Chap. 2)
α	Ratio between higher and lower heating value for fuel (Chap. 6)
δ	Electric current density (Chap. 7)
ε	Thermal emissivity (Chap. 8)
ε	Expansibility factor (Chap. 2)
η	Efficiency
ρ	Mass density
ρ	Electrical resistivity of conducting material (Chap. 7)
σ	Stefan–Boltzmann constant
ϕ, RH	Relative humidity
Ω	Rotational speed
ω	Humidity ratio or specific humidity
$\cos \varphi$	Electric power factor

Institutions and Associations

ABMA	American Boiler Manufacturers Association
ASHRAE	American Society of Heating, Refrigerating and Air Conditioning Engineers
ASME	American Society of Mechanical Engineers
IEC	International Electrotechnical Commission
IEEE	The Institute of Electrical and Electronics Engineers
ISO	International Organization of Standardization
TEMA	Tubular Exchanger Manufacturers Association

Introduction

1.1 General Principles of Energy Conversion and Management

In spite of a great diversity among energy end user technologies, due to technical, economic, and environmental factors, energy conversion and management at all levels must be based on a few general principles of proven validity.

Basically, every energy conversion must be performed by reducing related losses at each step, so as to improve the overall energy efficiency.

At the same time, energy management means ensuring that users get all the necessary energy, when and where it is needed, and of the quality requested, supplied at the lowest cost. Of course, all these aims must be achieved while duly safeguarding both production and environmental needs.

To be effective, energy management programs should include four main steps: (1) analyses of historical data, (2) energy audits and accounting, (3) engineering analyses and investment proposals based on feasibility studies, and (4) personnel training and information.

An energy management program can be organized in several ways, by employing either internal or external consultants, according to the company's size and the incidence of energy costs on the company's budget. In order to control energy conversion efficiency at every stage, a few significant energy key performance indexes must be identified before and after undertaking any energy-saving actions.

To obtain the best results, the main steps listed above must be carefully implemented and correlated with one another.

Energy management began to be considered as one of the main functions of industrial management in the 1970s. Faced with the rising price of energy and reports about the approaching exhaustion of world energy resources, both national governments and private companies had to cope with this situation with no further delay. In later years the correlation between energy consumption and environmental impact became evident, thus making people worldwide increasingly aware of the matter.

G. Petrecca, *Energy Conversion and Management: Principles and Applications*,
DOI 10.1007/978-3-319-06560-1_1, © Springer International Publishing Switzerland 2014

Inadequate knowledge, however, of energy management techniques and the lack of strong traditions in this kind of capital investment weighed heavily, at the beginning, against a widespread introduction of energy-saving strategies.

Large plants with high energy consumption tackled the problem by retrofitting process plants and facilities. Other industrial sectors, less sensitive to energy, resorted to investments with the shortest possible payback such as heat recovery and reduction of losses, but they put off process modifications, which often involve a more drastic change of production strategies.

The energy shortage in 1980–1990, the first that seriously affected the industrial era, and the following environmental impacts and oscillations in energy prices have made people aware that the energy problem does and will always exist. Therefore, energy-saving technologies have markedly influenced both component and plant designs, and users have become increasingly accustomed to making decisions on energy with regard to more than mere payback considerations.

Large plants and buildings seek energy managers with a good knowledge of technical and economic disciplines, since they play an important role in developing industrial strategies. Mechanics and thermodynamics have traditionally been the base in this educational field; but the spread of information technologies and power electronic systems suggests that electrical and thermodynamics approaches need to be interconnected and that technicians must be prepared to work within this broader context.

University students should have the opportunity to prepare themselves for these jobs; people already working in industry should try to integrate practical knowledge with basic theoretical principles in order to derive the maximum benefit from previous experience. Managers, who do not have time to study technical problems in detail, need a guide to the essentials in order to make well-founded decisions.

The aim of this book is to give an overall view of energy conversion and management by following the stream of energy from site boundaries to end users.

The author's philosophy is that most energy conversion problems can be approached through understanding the common principles underlying the phenomena.

All the topics are reduced to their essentials by introducing a few basic formulas and data. Basic KPIs, mainly drawn from the author's consultancy experience, are also reported. The tables in Chap. 20 highlight this approach and give a guidance to help understand and solve any problem by reducing it to a basic formula.

1.2 Energy Transformations in Factories and Buildings

The overall approach to energy management is summarized in Fig. 1.1 which shows the energy streams from a site's boundaries to end users and the related transformations:

- Energy is transported to the site as purchased fuel or utilities (electric network, gas pipeline, water pipe, etc.) and it is transformed mostly on site (electrical substation with transformers, boiler plants, cogeneration and trigeneration plants) before reaching the energy users. The energy requested by the end users can also be produced on site by means of renewable sources like sun and wind.

  Further transformations in facilities and plants should be made in order to obtain different forms of derived energy suitable for the end users.

  Obviously, checking the efficiency of all the transformation plants and keeping it as high as possible is a top priority.

- Energy streams in different forms (electric energy at different voltages, steam, hot water, chilled water, hot oil, compressed air, etc.) are distributed around the site to process and facility end users. Distribution systems are responsible for losses, which should be reduced by means of correct planning and thermal insulation.

- Energy end users around the site perform different operations which lead to end products or services. These make up the output of the site, together with waste (which may or may not contain energy in some form) and wasted energy.

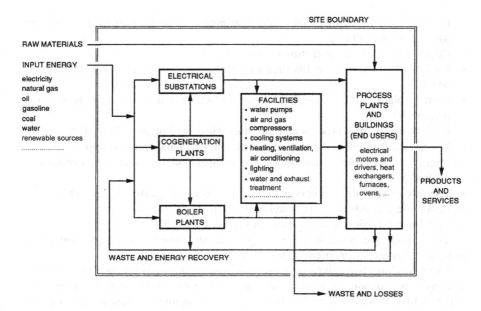

Fig. 1.1 Energy flow through the site boundary

Waste and wasted energy include water, solid and liquid materials (combustible or not), and gases.

Energy can be saved at the end user stage in many ways: by introducing more efficient equipment or process systems, by improving controls, by improving monitoring or metering of energy streams, and by recovering heat and waste. Waste recovery is closely related to waste quality and to the pollution produced.

Energy flow which crosses the site boundaries can be schematized as in Fig. 1.1. This energy flow chart allows an overall approach to site energy management and provides the guidelines for an energy survey as well as an educational program.

1.3   The Plan of the Book

The book has been conceived both as a textbook for university courses in engineering and as a work of reference for professionals in energy management. Readers are assumed to have a basic knowledge of thermodynamics, heat and mass transfer, electric systems and power electronics, as well as computer programming.

The book comprises 20 chapters that can be grouped as follows:
- General principles of energy transformation, energy management, and energy sources (Chaps. 1–4):

    This part aims to give readers a general overview of energy conversion and management, from the energy sources to the end users.
- Transformation plants at the site borders (Chaps. 5 and 6): Electrical substations and boiler plants are examined and suggestions are made on how to improve efficiency.
- Internal electric networks and fluid distribution systems from facilities to end users (Chaps. 7 and 8).
- Cogeneration and trigeneration plants (Chap. 9).
- Facility plants to move fluids such as pumps, fans, and compressors (Chaps. 10 and 11).
- Facility plants such as cooling plants, HVAC systems, and lighting systems (Chaps. 12–14).
- Heat recovery from process and facilities: Heat exchangers (Chap. 15).
- Waste management and correlation with energy management (Chap. 16).
- Energy audits, energy accounting for control and planning, and centralized control (Chap. 17).
- The role of education in energy conversion and management (Chap. 18).
- Economic analysis of energy-saving investments (Chap. 19).
- Conclusion: Basic formulas, data, and KPIs (Chap. 20).

Throughout the book boxes highlight the fundamental concepts and some important figures for practical applications.

Examples have been given only for basic cases and mainly for facilities, but they can easily be extended to more complicated situations, including process systems, which at first glance seem quite different from one another, if these are not reduced to their essentials. Technical evaluations are shown at the end of each chapter; data is generally organized in tables to facilitate elaboration by a standard spreadsheet. Economic evaluations of each example are shown in Table 19.3.

Basic formulas, data, and KPIs that everybody involved in energy conversion should know are reported in Chap. 20.

The SI system has been used; for ease of understanding, other units commonly used have been added.

Efforts have been made to attribute a single-letter symbol to each physical quantity. Because of the number of topics introduced, some symbols have different meanings according to the context and to the common practice; in these cases, the list of symbols indicates the chapters to which these symbols apply.

References are at the end of the book in chronological order.

Power and Energy Measurement Units and Techniques

2

2.1 The SI System and Conversion Factors

The SI system (International System of Units) is a set of definitions and rules which provides an organic and consistent approach for obtaining the units for each quantity.

The SI system is based on seven basic and two supplementary units. The basic units are the following: the kilogram (mass), the meter (length), the second (time), the ampere (electric current), the kelvin (temperature), the candela (luminous intensity), and the mole (molecular substance). The supplementary units are the radian (plane angle) and the steradian (solid angle). The choice, which has been made by the General Conference on Weights and Measures, is always open to changes and integrations.

The SI basic and supplementary units are reported in Table 2.1. From these nine units any other SI units can be derived by using the following expression:

$$\text{Derived unit} = m^{\alpha_1} \times kg^{\alpha_2} \times s^{\alpha_3} \times A^{\alpha_4} \times K^{\alpha_5} \times cd^{\alpha_6} \times mol^{\alpha_7} \times rad^{\alpha_8} \times sr^{\alpha_9}$$

where the exponents α_1, α_2, α_3, α_4, α_5, α_6, α_7, α_8, α_9 are positive or negative whole numbers. Each exponent is equal to zero when the derived unit does not depend on the corresponding basic or supplementary unit.

Table 2.2 shows the prefixes which must be used to derive multiples or submultiples of any SI unit. Table 2.3 shows a few derived units currently used.

It is worth noticing that EU (European Union) countries must use the SI units reported in Table 2.1 (they are obliged to by European Commission's recommendations and related national laws). This system has been adopted also by the International Organization of Standardization (ISO).

Table 2.4 lists commonly used units which, however, do not belong to the SI system. Technicians should refrain from using these units.

The table shows the conversion factors from these units to SI units; conversion between two non-SI units can be made by using the ratio between the conversion

G. Petrecca, *Energy Conversion and Management: Principles and Applications*, DOI 10.1007/978-3-319-06560-1_2, © Springer International Publishing Switzerland 2014

Table 2.1 SI base and supplementary units

Quantity	Unit	Symbol
Length	meter	m
Mass	kilogram	kg
Time	second	s
Electric current	Ampere	A
Thermodynamic temperature	Kelvin	K
Luminous intensity	candela	cd
Molecular substance	mole	mol
Plane angle	radian	rad
Solid angle	steradian	sr

Table 2.2 Prefixes commonly used

Factor	Prefix name	Symbol
Multiple		
10^{18}	Exa	E
10^{15}	Peta	P
10^{12}	Tera	T
10^{9}	Giga	G
10^{6}	Mega	M
10^{3}	kilo	k
10^{2}	hecto	h
10^{1}	deka	da
Submultiple		
10^{-1}	deci	d
10^{-2}	cents	c
10^{-3}	milli	m
10^{-6}	micro	μ
10^{-9}	nano	n
10^{-12}	pico	p
10^{-15}	femto	f
10^{-18}	atto	a

factors of the single units. This is valid also for temperature, if temperature changes are considered (as often occurs in many formulas).

Table 2.5 lists some parameters frequently used for technical calculations.

2.2 Primary Energy Measurement Units

The primary energy content of combustibles (solid, liquid, or gaseous) is expressed by Lower Heating Values or Higher Heating Values such as kJ/kg or kJ/m³ (SI units). There are other units in common use.

Table 2.3 Units derived from SI

Quantity	Unit	Symbol
Space and time		
Area	square meter	m^2
Volume	cubic meter	m^3
Velocity	meter per second	m/s
Acceleration	meter per second squared	m/s^2
Angular velocity	radian per second	rad/s
Angular acceleration	radian per second squared	rad/s^2
Frequency	Hertz	$Hz = cycle/s$
Mechanics		
Density	kilogram per cubic meter	kg/m^3
Momentum	kilogram meter per second	$kg \cdot m/s$
Moment of inertia	kilogram meter squared	$kg \cdot m^2$
Force	Newton	$N = kg \cdot m/s^2$
Torque, moment of force	Newton meter	$N \cdot m$
Energy, work, heat quantity	Joule	$J = N \cdot m$
Power	Watt	$W = J/s$
Pressure, stress	Pascal	$Pa = N/m^2$
Electricity and magnetism		
Electric charge	Coulomb	$C = A \cdot s$
Electric potential, voltage	Volt	$V = W/A$
Electric field strength	Volt per meter	V/m
Capacitance	Farad	$F = C/V = A\ s/V$
Current density	Ampere per square meter	A/m^2
Magnetic field strength	Ampere per meter	A/m
Magnetic flux	Weber	$Wb = V \cdot s$
Magnetic flux density	Tesla	$T = Wb/m^2$
Inductance	Henry	$H = V \cdot s/A$
Permeability	Henry per meter	H/m
Resistance	Ohm	$\Omega = V/A$
Conductance	Siemens	$S = A/V$
Magnetomotive force	Ampere	A
Light		
Luminous flux	lumen	$Im = cd \cdot sr$
Illuminance	lux	$lx = 1\ m/m^2$
Viscosity		
Kinematic viscosity	square meter per second	m^2/s
Dynamic viscosity	Pascal second	$Pa \cdot s$

Table 2.4 SI units and conversion factors

To convert from	Symbol	To	Symbol	Multiply by
Length				
foot	ft	meter	m	0.3048
inch	in	meter	m	0.0254
mile	mi	meter	m	1,609
Area				
square foot	ft^2	square meter	m^2	0.0929
square inch	in^2	square meter	m^2	0.004645
Volume				
cubic foot	ft^3	cubic meter	m^3	0.02832
cubic inch	in^3	cubic meter	m^3	0.00001639
USA liq gallon	gal	cubic meter	m^3	0.0037854
Liter	L	cubic meter	m^3	0.001
Mass				
pound	lb	kilogram	kg	0.45359
ton(short)	ton	metric ton, tonne	$t = 10^3$ kg	0.9072
ton(long)	ton	metric ton, tonne	$t = 10^3$ kg	1.016
barrel(oil)	barrel	metric ton, tonne	$t = 10^3$ kg	0.137
Force				
pound-force	lbf	Newton	N	4.448
kilogram-force	kgf	Newton	N	9.807
Pressure				
pound-force/ square foot	lbf/ft^2	Pascal	Pa	47.8788
pound-force/ square inch	lbf/in^2	Pascal	Pa	6,895
kilogram-force/ square meter	kgf/m^2	Pascal	Pa	9.807
bar	bar	Pascal	Pa	100,000
atmosphere	atm	Pascal	Pa	101,325
mm H_2O	mm H_2O	Pascal	Pa	9.7739
inch H_2O	in H_2O	Pascal	Pa	248.7
Speed, velocity				
foot/second	ft/s	meter/second	m/s	0.3048
foot/min	ft/min	meter/second	m/s	0.00508
mile/hour	mi/h	meter/second	m/s	0.4469
kilometer/hour	km/h	meter/second	m/s	0.2777
Acceleration				
foot/second^2	ft/s^2	meter/second	m/s^2	0.3048
Energy, work				
British thermal unit	Btu	Joule	J	1,055
foot pound-force	ft lbf	Joule	J	1.356
calorie	cal	Joule	J	4.1868
Watthour	Wh	Joule	J	3,600
Power				
Btu/hour	Btu/h	Watt	W	0.2931
Btu/second	Btu/s	Watt	W	1,055
horsepower	hp	Watt	W	745.7
calorie/hour	cal/h	Watt	W	0.0011628

(continued)

Table 2.4 (continued)

To convert from	Symbol	To	Symbol	Multiply by
Refrigerant				
capacity tons	tons	Watt	W	3,520
frigorie/hour	frig/h	Watt	W	0.0011628
Torque				
pound-force foot	lbf ft	Newton meter	N·m	1.356
kilogram-force meter	kgf m	Newton meter	N·m	9.807
Density				
pound/cubic foot	lb/ft^3	kilogram per cubic meter	kg/m^3	16.018
Volume flow rate				
cubic foot/minute	ft^3/min	cubic meter per second	m^3/s	0.00047
Specific energy				
Btu/pound	Btu/lb	Joule/kilogram	J/kg	2,326
calorie/kilogram	cal/kg	Joule/kilogram	J/kg	4.186
Specific heat				
Btu/pound °F	Btu/lb °F	Joule/kilogram K	J/kg K	4.186
calorie/kilogram °C	cal/kg °C	Joule/kilogram K	J/kg K	4.186
Light				
footcandle	fc	lux	lx	10.764
Temperature				
Celsius °C change	°C	Kelvin change	K	1
Fahrenheit °F change	°F	Kelvin change	K	5/9

Note that conversion between two non SI units can be made by using the ratio between the conversion factors of the single unit

Examples

To convert from	To	Multiply by
Celsius change	Fahrenheit change	1/(5/9)
Btu/h	cal/h	0.2931/0.0011628

The quantity of combustibles is generally referred to TOE (Ton Oil Equivalent) by using the Lower Heating Value (41,860 kJ/kg).

Primary energy (hydro, geothermal, nuclear, or other renewable energy sources) from which electric energy is produced is converted into TOE on the basis of the specific consumption (kJ/kWh) of conventional fuel-fed utility power plants. Typical values range between 6,600 and 10,500 kJ/kWh (corresponding to 6,256 and 9,952 Btu/kWh) if both power plant and distribution line losses are taken into account; lower values correspond to combined cycles with gas turbines, whereas higher values to boilers and steam condensing turbines fed by solid coal.

Table 2.5 Parameters frequently used

Description	Other systems		SI system
Specific heat	kcal/kg °C	Btu/lb °F	kJ/kg K
Water	1	1	4.18
Superheated steam[a]	0.5	0.5	2.09
Air	0.24	0.24	1
Iron	0.114	0.114	0.477
Copper	0.092	0.092	0.385
Mineral oil	0.486	0.486	2.034
Density[b]		lb/ft^3	kg/m^3
Water		62.5	1,000
Air (standard conditions)		0.08	1.29
Mineral oil		57.75	925
Iron		490	7,850
Copper		557.5	8,930
Natural gas		0.047	0.750

[a]Average value in industrial boiler. In air-water mixture (see Chap. 13) the specific heat of superheated steam equals 1.8 kJ/kg K (steam pressure <0.1 MPa)
[b]Density is referred to standard conditions: 0.1 MPa (1.013 bar, 14.5 psi), 273.15 K (0 °C; 32 °F) for air, and 288.75 K (15.6 °C; 60 °F) for natural gas
For an ideal gas, but widely accepted for most real gases, basic relationships (where V is the volume) are:

$$pV = \text{constant} \cdot T$$

The density of air at T_1 (K) and at standard pressure is: air density $(T_1) = \dfrac{1.293}{\left(\frac{T_1}{273.15}\right)}$

Notice that in some countries and in some applications the standard conditions can be different from the previous ones. For natural gas: 288.75 K (15.6 °C; 60 °F) and 0.1 MPa (14.5 psi). The user should ascertain the reference conditions for each application

Tables 2.6 and 2.7 report values from international statistics and conversion factors.
Some definitions, which will be discussed in detail in later chapters, are summarized below in order to allow a clearer understanding of Chap. 2:
• Heating value. This is a measure of the heat any given fuel can release during the combustion process.

Combustion of fuels consisting of carbon, hydrogen, and sulfur requires oxygen, which normally comes from atmospheric air. It starts at different ignition temperatures depending on the fuel. Typical values range between 573.15 and 973.15 K (300–700 °C, 572–1,292 °F).

Once ignition temperature has been reached, combustion continues until all the fuel or oxygen has been consumed.

Hydrogen, combined with oxygen, produces water (roughly 9 kg of water for 1 kg of hydrogen), which is discharged as liquid water as well as water vapor into the atmosphere together with combustion gaseous waste at the same temperature;
• Higher Heating Value (HHV), also called Gross Heating value. This is the number of heat units measured as being liberated when unit mass of fuel is burned in oxygen saturated with water vapor in a bomb in standard conditions, the residual materials being gaseous oxygen, carbon dioxide, sulfur dioxide and

Table 2.6 Lower and Higher Heating Values of solid, liquid, and gaseous fuels

Fuels	Average values			
	Lower	Higher	Lower	Higher
Solid fuels	*kJ/kg*		*Btu/lb*	
Vegetal fuels	10,465	16,700	4,499	7,180
Pitch lignite	18,000	24,000	7,738	15,478
Standard coal	29,302	33,500	12,598	14,403
Charcoal	31,395	34,750	13,498	14,941
Cokery coke	29,302	33,000	12,598	14,188
Gas coke	26,790	32,650	11,518	14,038
Petroleum coke	34,744	37,250	14,938	16,015
Liquid fuels	*kJ/kg*		*Btu/lb*	
Crude oil	41,860	44,400	17,997	19,090
Petroleum condensates	44,372	47,000	19,077	20,207
Light petroleum distillates	43,534	46,150	18,717	19,842
Gasoline	43,953	46,600	18,897	20,035
Jet fuel	43,534	46,150	18,717	19,842
Refined kerosene	43,116	45,700	18,537	19,648
Gasoil	42,697	45,250	18,357	19,455
Fuel oil	41,023	43,500	17,637	18,702
Liquid hydrogen	120,070	141,800	51,621	60,964
Propane	46,296	50,235	19,904	21,597
Liquefied petroleum gas (LPG)	46,046	49,700	19,797	21,368
Gaseous fuels	*kJ/m^{3a}*		*Btu/ft^{3a}*	
Natural gas	34,325	38,450	921	1,032
Methane	34,285	38,000	953	1,057
Cokery gas	17,791	19,900	478	534
Blast furnace gas	3,767	4,200	101	113

[a]0.1 MPa (14.5 psi), 288.75 K (15.6 °C; 60 °F) for natural gas and 273.15 K (0 °C; 32 °F) for other gases

Table 2.7 Conventional densities of liquid fuels

Fuels	kg/m^{3a}	ib/ft^{3a}
Gasoline	734	45.8
Gasoil	825	51.5
Oil	925	57.7
LPG	565	35.3
Natural gas[a]	0.75	0.047
Standard coal	800	43.9
Vegetal fuels	400	24.3

[a]0.1 MPa (14.5 psi), 288.75 K (15.6 °C; 60 °F) for natural gas

nitrogen, ash, and liquid water (the water produced during the combustion is assumed to be discharged as liquid water). The standard conditions are defined by ISO. The international reference temperature for combustion is 25 °C (77 °F), but in some countries different temperatures are used;

- Lower Heating Value (LHV), also called Net Heating Value. This is the number of heat units measured as before, the residual materials being gaseous oxygen, carbon dioxide, sulfur dioxide and nitrogen, ash, and water vapor (the water produced during the combustion is assumed to be discharged as water vapor). The water vapor enthalpy, which is completely wasted, is not taken into consideration.

2.3   End User Energy Measurement Units

The energy entering a site goes through many transformations before reaching end users (this energy is commonly called final energy). Units referring to transformed energy rather than to purchased energy (fuels and electric energy) are conveniently used at this stage. Table 2.8 lists the typical ranges of values for the commonest ratios between end user and purchased energy, which depend on the efficiency of the transformations as well as on the level of energy exploitation:

- kg steam/kg fuel;
- kg hot water/kg fuel;
- cubic meter compressed air/kWh consumed by mechanical compressor;
- kJ available for the process/kWh consumed by mechanical compressor;
- Others.

Table 2.8 Ratios between end users energy and purchased energy (see also Chaps. 18 and 20)

Types of transformation				
To convert from	To	Transformation coefficient	Unit	Note
Oil	Steam	12–14	kgsteam/kgoil	Boiler
Natural gas	Steam	9–11	kgsteam/Sm^3	Boiler
Electricity	Cold fluid (HVAC)	12,000–16,000	kJ/kWh	Compressor
Electricity	Cold fluid (below ice point)	3,000–10,000	kJ/kWh	Compressor
Electricity	Compressed air (0.8 MPa)	9–10	Sm^3/kWh	Compressor
Electricity	Heat	3,600	kJ/kWh	Resistor
Electricity	Heat	10,000–16,000	kJ/kWh	Heat pump
Oil	Electricity	4–7	kWh/kg	Utility plant
Oil	Electricity	8.5–9.5	kWh/kg	Cogeneration plant
Electricity	Water storage	250–300	t · m/kWh	Pump
Electricity	Lighting	50–100	lm/W	Lamps

2.4   An Outline of the Main Measurement Techniques

Energy measurement is important because on the one hand it is a means of improving energy management and control of plant operations and on the other it determines the incidence of energy cost on the total cost of production.

Energy can be found in many different forms:
- Solid, liquid, and gaseous fuels;
- Electric energy;
- Fluids by which heat is distributed throughout a site, such as steam, hot water, hot oil, heated air, etc.;
- Fluids by which motive power is distributed throughout a site, such as compressed air, compressed oil, etc.

All these energies can be measured by means of two categories of system:
- Meters that measure energy flow over time by performing the integral function inside the instrument;
- Transducers of instantaneous values such as flow and power. Energy is then calculated by means of independent systems able to perform an integral function. Notice that it is also possible to calculate energy if physical parameters such as temperature, pressure, etc. are measured and models of the phenomena are known.

2.4.1   Liquid and Gas Measurements

Flow meters for combustibles and other fluids, both liquids and gas, can be grouped in volume meter; velocity meter; head meter; pitot tube; vortex; rotameter; sonic meter; magnetic meter; Coriolis and thermal mass flow meters; others.

Notice that these meters always measure volume flow rate at actual temperature and pressure. In the case of gas, the meter indication must be corrected, generally inside the measurement system, in order to obtain volume flow rate in standard conditions by taking into account pressure and temperature as follows:

$$q_s = q \times \frac{p}{p_s} \times \frac{T_s}{T} \times \frac{Z_s}{Z}$$

where
q_s = standard condition volume flow rate (Sm^3/s)
q = meter indication actual volume flow rate (m^3/s)
p_s = standard atmospheric pressure (0.1013 MPa)
p = absolute actual pressure (MPa)

(continued)

(continued)

T_s = standard temperature (273.15 K)
T = actual gas temperature (K)
Z_s, Z = compressibility factors.

Mass flow rate is then obtained:

$$m \ (\text{kg/s}) = q_s \times \rho_s$$

where
ρ_s = mass density at standard conditions (kg/Sm^3).

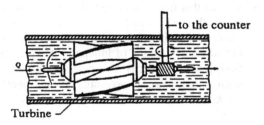

Fig. 2.1 Velocity meter with a turbine rotor

1. Volume meter. Volume meters generally operate on the principle of giving an indication proportional to the quantity that has flowed into fixed-volume containers over time. With alternate filling and emptying of the container, an indication of how many times the container is filled and emptied can easily be transferred to a counter with a calibrated dial in order to determine flow quantity.

    Gasoline and gas meters were widely used in the past, but now they are outdated; other types of volume meters are still in use.

    The error ranges from 0.5 % to 2.5 % depending on the meter model; a good degree of accuracy can be obtained from full capacity down to zero (working range 100:1).

    Notice that volume meters can be installed anywhere along the pipe, but particular attention must be paid to floating solid matter. Filters must be installed and they should be cleaned on a routine basis.

2. Velocity meter. The quantity or the volume of the fluid is derived by means of a velocity measure that is transferred to a counter, generally calibrated in rates of total flow.

    The operating principle is as follows: the meter has an orifice with a shunt circuit around it. Fluid flows through it and compels a turbine or a helix to rotate (see Fig. 2.1). The rotation speed, which is proportional to the fluid velocity, is transmitted to a counter calibrated in rates of flow through the orifice (volume flow rate = velocity × area).

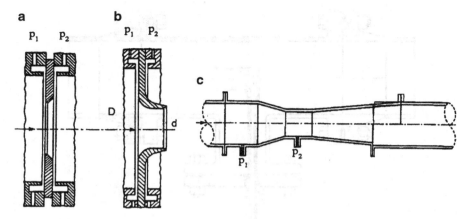

Fig. 2.2 Different types of head meters. (**a**) Thin plate orifice; (**b**) sharp edge orifice; (**c**) Venturi tube. Upstream (p_1) and downstream (p_2)

The meter can be installed anywhere along the pipe; in bypass installation, the counter is calibrated by taking into account the diameters of the orifice and the nozzle of the main pipe.

Overloads up to 150–200 % of rated capacity are handled temporarily without loss of accuracy.

Both volume and velocity meters are able to work with liquid and gas, but they are not interchangeable (for instance, if gas flows through a meter originally designed for a liquid, the rotor velocity will be too high and mechanical damage may occur).

3. Head meter. The operating principle is based on the measurement of a differential pressure (pressure losses) between two points of the pipe; this is produced by introducing an orifice or a nozzle. In this way fluid speed increases and pressure decreases.

The differential pressure between the two points is proportional to the square of the mass flow rate, thus:

$$q = \alpha \times S \times \sqrt{(p_1 - p_2)/\rho}$$

where q = volume flow rate (m^3/s), α = instrument flow coefficient (expansibility factor for gas included), S = section of the orifice or the nozzle (m^2), ρ = fluid mass density (kg/m^3), p_1, p_2 = upstream and downstream pressures (Pa).

Head meters differ with the type of device producing a differential pressure:

• Thin plate and sharp edge orifices (see Fig. 2.2a, b). An orifice with a diameter d is inserted by means of two shafts anywhere along the pipe with a diameter D. Values of the diameter ratio d/D ranging from 0.05 to 0.7 and straight pipes upstream and downstream are required. The length of the straight pipes must be at least $20D$ upstream and $5D$ downstream;

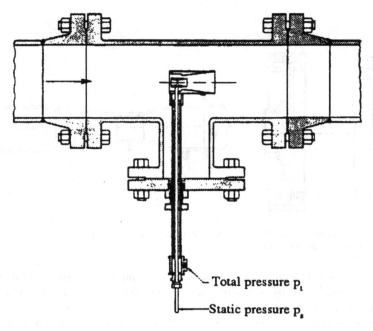

Fig. 2.3 Pitot tube

- Venturi tube (see Fig. 2.2c). This meter uses a Venturi nozzle which permits
  flow measurement with low differential pressure values. Then shorter straight
  pipes upstream and downstream are required, generally less than 15D.

  Notice that the error of head meters is generally less than 2 %, but increases as
  the meter works in the lowest part of the scale because of the squared relationship between flow and differential pressure. Care must be taken not to oversize
  the full-scale capacity on the basis of the real flow rate; a good degree of
  accuracy is obtained from full capacity down to 25 % (working range 4:1).

4. Pitot tube (see Fig. 2.3). This operates on the principle that the total pressure p_t,
   detected at a small orifice facing into the flow, is the sum of static and dynamic
   pressures. The static pressure p_s can be detected by a measurement orthogonal to
   the direction of the flow. The dynamic pressure is then obtained by subtraction.
   Notice that straight pipes are required upstream and downstream in order to
   avoid turbulence near the point of measurement.

   Fluids without solid matter are required. Flow speed must be relatively high,
   because low speed involves a differential pressure too slight to measure easily.
   A squared relationship exists between flow and differential pressure, as in head
   meters.

   The basic operating formula is as follows:

$$q = \alpha \times K_v \times S \times \sqrt{(p_t - p_s)/\rho}$$

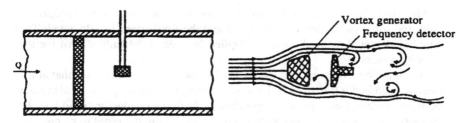

Fig. 2.4 Vortex meter

where q = volume flow rate (m³/s), α = instrument flow coefficient (expansibility factor for gas included), S = section of the orifice or the nozzle (m²), K_v = velocity coefficient which depends on the Reynolds number, ρ = fluid mass density (kg/m³), p_t, p_s = total and static pressures (Pa).

5. Vortex (see Fig. 2.4). The operating principle is based on Karman's law, which concerns vortices produced by a fluid. A vortex generator, which is an appropriately shaped solid, is placed inside the pipe and the fluid vortex frequency is detected. The working equation is as follows:

$$q = S \times V = S \times (K_m \times f)$$

where q = volume flow rate (m³/s), S = section of the pipe (m²), V = fluid velocity (m/s) which is equal to the vortex frequency f (Hz) multiplied by the vortex meter constant (K_m = Karman constant).

Vortex meters are able to measure volume flow rates (mass flow rates are derived by introducing the volume flow rate at standard condition and the related mass density) of gas, liquid, and vapor. A good degree of accuracy is obtained from full capacity down to 10 % (working range 1:10). In the case of gas, a correction factor as with velocity meters is required. It is generally elaborated inside the measurement system to give the value of the standard volume flow rate and the real value of the mass flow rate.

6. Rotameter. This operates on the principle that a flowing fluid exerts a force proportional to the flow. A rotameter is a vertical conic tube in which a ball or any other calibrated device made of plastic or glass moves freely. The calibrated device is subject to two opposite forces: the force of gravity which works downwards and the fluid force which works upwards. The device moves vertically and indicates the rate of the fluid flow.

Notice that no straight pipes upstream or downstream are required; the only condition is that fluid must flow upwards from below.

The rotameter is suitable for measuring low rates of flow and must be differently calibrated for each kind of fluid.

7. Sonic and magnetic meter. The operating principle of a sonic meter is based on the fluid property of propagating a sound at a speed which depends on fluid volume, pressure, and temperature.

Electric pulses are generated and transformed into sonic ones through the fluid; the sonic pulses are then detected and the fluid speed is calculated. The flow is calculated as fluid speed multiplied by area, as already shown for the velocity meters.

A magnetic meter works on the basis of Faraday's law which says that at the ends of a conductor (the fluid in this case) moving inside a magnetic field there is an electromotive force linearly depending on both the conductor speed and the magnetic field itself. This meter, which does not require the pipes to be cut, can be used only with liquid with a conductivity of not less than 500 μS/mm.

8. Coriolis mass meter, thermal mass meter. Coriolis meter is based on the controlled generation of Coriolis forces inside parallel tubes, where vibrations (ω) are detected. The mass flow rate is proportional to ω^2. In a thermal mass flow rate the mass flow is directly proportional to the rate of heat absorbed by the flow stream.

9. Other systems based on different principles are also available but it is not necessary to describe all of them in detail. They can readily be found in specialized books.

Table 2.9 shows basic operating parameters able to qualify flow meters and their applications. Note that the commonest devices are head meters, with orifice or Venturi tube, Pitot tube meters, vortex meters, and thermal mass flow meters. The choice among them depends on the fluid to be measured and on the accuracy required.

2.4.2   Electric Energy Flow Measurements

Equipment for measuring energy flow and power demand in the distribution system consists of conventional voltmeters, ammeters, and kilowatthour meters. Potential transformers (PT) or current transformers (CT) are generally installed to reduce the level of voltages and currents for technical as well as for security reasons. Current is generally stepped down to 5 A or less, line voltage to 120 V or 100 V or less.

Specific equipment, which performs combinations of the abovementioned functions, is also available to provide kilovarhour, kilovoltamperehour, kilowatt, kilovoltampere, and power factor.

Typical connection schemes of the three basic meters are shown in Fig. 2.5, where basic relationships between current, voltage, and power are also illustrated. Notice that:

- Voltmeter is a high-resistance coil device and it must always be connected in parallel;
- Ammeter is a low-resistance coil device and it must always be connected in series with the current. With portable meters, clamp-around current transformers are generally used;
- Wattmeter, which measures real power, reactive or apparent power, is a combination of potential coils (parallel connected) and current coils (series connected);

Table 2.9 A review of the main flow meters

Type of device	Pipe diameter (mm)	Type of fluid to measure			Max pressure		Max temperature		Reynolds number	Operating range	Typical accuracy (%)	Pressure losses (kPa)
		Liquid	Gas	Vapor	MPa	psi	°C	°F				
Volume meter	5–500	x	x	–	20	2,900	100	212	Any value	50:1	0.2–0.5	50–100
Velocity meter	10–600	x	x	–	10	1,450	200	392	>10,000	15:1	0.2–0.5	5–10
Head meters: orifice and Venturi tube	>50	x	x	x	50	7,252	500	932	>2,500	4:1	1–2	2,050
Pitot	>100	x	x	x	50	7,252	500	932	Any value	4:1	2–5	1–2
Vortex	25–200	x	x	x	10	1,450	200	392	>10,000	15:1	0.75–1.5	10–20
Rotameter	5–150	x	x	x	4	580	200	392	>10,000	10:1	2–3	5–10
Sonic meter	Any value	x	x	–	20	2,900	200	392	Any value	10:1	2–3	0
Magnetic meter	Any value	x	–	–	20	2,900	200	392	Any value	20:1	0.5	0
Coriolis mass meter	5–100	x	x	–	30	4,351	150	302	Any value	100:1	0.1–0.5	10–20
Thermal mass meter	>5	x	x	–	10	1,450	200	392	Any value	100:1	0.5–2	10–20

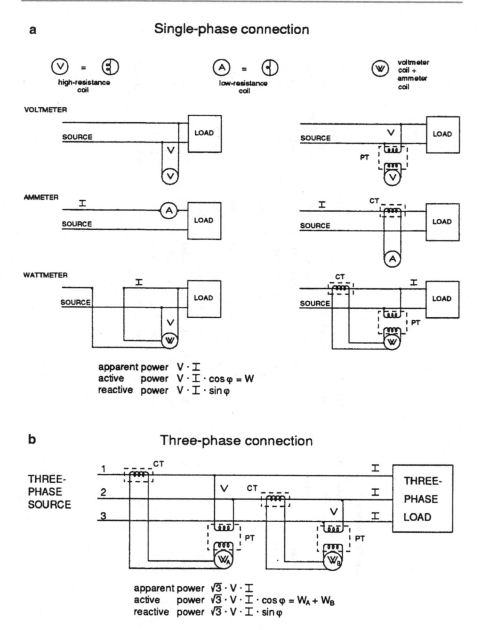

a   Single-phase connection

apparent power $V \cdot I$
active   power $V \cdot I \cdot \cos\varphi = W$
reactive power $V \cdot I \cdot \sin\varphi$

b   Three-phase connection

apparent power $\sqrt{3} \cdot V \cdot I$
active   power $\sqrt{3} \cdot V \cdot I \cdot \cos\varphi = W_A + W_B$
reactive power $\sqrt{3} \cdot V \cdot I \cdot \sin\varphi$

Fig. 2.5 Typical connection schemes: voltmeter, ammeter and wattmeter with single-phase line (**a**) and with three-phase line (**b**)

- When current transformers or potential transformers are used, the CT and the PT ratios must be taken into account. The meter readings multiplied by the meter constant and the CT or PT ratios (or both in the case of wattmeter) give the values of current, voltage, and power.

  Typical CT ratios are 100:5, 1,000:5, 5,000:5. These values show how many amperes flow in the primary conductor or line conductor when a 5 A current flows in the secondary winding. For security reasons the secondary winding must never work with no-load, and consequently it must always be connected to a low-resistance ammeter or shorted by means of a jumper wire, screw, or switch on a CT shorting terminal strip.
- Complex metering systems can also be introduced by using current and voltage transducers whose signals are elaborated by means of computers and transformed into electric power and energy. The introduction of computerized systems allows both the measurement of energy flow and the control of electric systems based on different optimization criteria.

  These systems can also be integrated conveniently with flow and other physical parameter measurements in order to obtain complete process control. This approach, quite common in energy-intensive industry, is becoming popular also in many manufacturing sectors and in buildings (see Sect. 17.8).

2.4.3 Temperature, Pressure, and Other Measurements

Devices for measuring temperatures can be grouped according to the principle of operation and the temperature range. The commonest devices are liquid-in-glass thermometers, bimetallic thermometers, resistance temperature devices, thermocouples, infrared thermometers, and optical pyrometers. A correct choice must be based on temperature ranges, environmental conditions, and uses of the measurements for reporting or process control.

Figure 2.6 shows relationships among the Kelvin (K), Celsius (°C), Rankine (°R), and Fahrenheit (°F) temperatures.

Devices for pressure measurement can be listed as Bourdon gage, diaphragm gage, and manometer. They measure the difference between the absolute pressure in a system and the absolute pressure of the atmosphere outside the measuring device. This difference is called differential or gage pressure. If the pressure of the system is lower than atmospheric pressure, the term vacuum pressure is used instead of gage pressure.

The Bourdon gage consists of a curved tube closed at one end with the other end connected to the pressure to be measured. When the pressure inside the tube is greater than the pressure outside, the tube tends to straighten and the amount of change in length or curvature can be translated into a gage reading.

The diaphragm gage is based on the detection of the diaphragm movement if the pressures against its two sides are different.

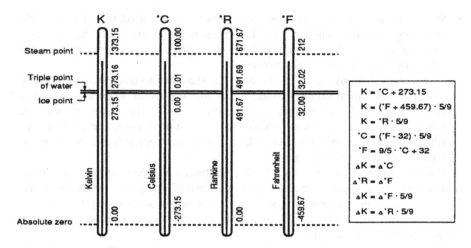

Fig. 2.6 Relationships among the Kelvin (K), Celsius (°C), Rankine (°R) and Fahrenheit (°F) temperatures

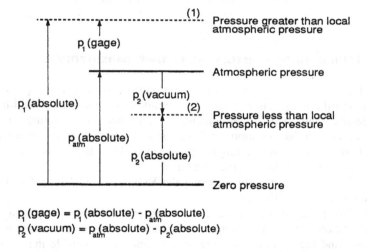

$$p_1 \text{ (gage)} = p_1 \text{ (absolute)} - p_{atm} \text{(absolute)}$$
$$p_2 \text{ (vacuum)} = p_{atm} \text{(absolute)} - p_2 \text{(absolute)}$$

Fig. 2.7 Relationships among the absolute, atmospheric, gage and vacuum pressures

The manometer is a tube with liquid in it; one end is open to the air and the second end is exposed to a different pressure, so that the end with the higher pressure has a lower liquid level and the pressure difference is detected.

Figure 2.7 shows relationships among the absolute, atmospheric, gage, and vacuum pressures.

Other physical parameter measurements can be made to meet specific needs. A careful study of specialized books and manufacturers' technical specifications is suggested in order to make the right choice for each application.

World Energy Demand

3

3.1 Introduction

Trends in energy use are expected to increase all over the world as the population is growing together with its need for goods and comfort.

The discovery of innovative forms of energy conversion on the earth and in outer space—not necessarily based on fossil fuels, nuclear, and the current renewable sources—as well as energy saving in industries, buildings, and transportation might contribute to smoothing both the increase of energy consumption per person and the strain of energy as a worldwide political factor.

3.2 World Energy Demand and Population

A reliable forecast of energy resources, energy consumption, and population in the future is a difficult task and past forecasts have failed, like those on the shortage of fossil fuels. So, instead of absolute figures about future energy demand and sources worldwide, which would become outdated within a short time, the matrix in Table 3.1 correlates the gross primary energy use per person (TOE/person) with the world population with the aim of providing a set of reliable figures at anytime.

At the beginning of the twenty-first century this value was between 0.8 and 5 TOE/person, depending on the level of both industrialization and urbanization, the latter strictly related to life comfort; 1.8 TOE/person can be assumed as a weighted average value at the beginning of year 2000 with a world population of 6,500 million. If we assumed the upper value (5 TOE/person) as a cap for the energy consumption of the world population in the future, the gross primary energy consumption range could be estimated on the basis of the population figures.

G. Petrecca, *Energy Conversion and Management: Principles and Applications*,
DOI 10.1007/978-3-319-06560-1_3, © Springer International Publishing Switzerland 2014

Table 3.1 Correlation matrix between population and primary energy consumption per person

Primary energy consumption per person (TOE/unit)	Total energy consumption (MTOE/year)											
Cap value—year 2000												
5.0	32,500	35,000	37,500	40,000	42,500	45,000	47,500	50,000	52,500	55,000	57,500	60,000
4.8	31,200	33,600	36,000	38,400	40,800	43,200	45,600	48,000	50,400	52,800	55,200	57,600
4.6	29,900	32,200	34,500	36,800	39,100	41,400	43,700	46,000	48,300	50,600	52,900	55,200
4.4	28,600	30,800	33,000	35,200	37,400	39,600	41,800	44,000	46,200	48,400	50,600	52,800
4.2	27,300	29,400	31,500	33,600	35,700	37,800	39,900	42,000	44,100	46,200	48,300	50,400
4.0	26,000	28,000	30,000	32,000	34,000	36,000	38,000	40,000	42,000	44,000	46,000	48,000
3.8	24,700	26,600	28,500	30,400	32,300	34,200	36,100	38,000	39,900	41,800	43,700	45,600
3.6	23,400	25,200	27,000	28,800	30,600	32,400	34,200	36,000	37,800	39,600	41,400	43,200
3.4	22,100	23,800	25,500	27,200	28,900	30,600	32,300	34,000	35,700	37,400	39,100	40,800
3.2	20,800	22,400	24,000	25,600	27,200	28,800	30,400	32,000	33,600	35,200	36,800	38,400
3.0	19,500	21,000	22,500	24,000	25,500	27,000	28,500	30,000	31,500	33,000	34,500	36,000
2.8	18,200	19,600	21,000	22,400	23,800	25,200	26,600	28,000	29,400	30,800	32,200	33,600
2.6	16,900	18,200	19,500	20,800	22,100	23,400	24,700	26,000	27,300	28,600	29,900	31,200
2.4	15,600	16,800	18,000	19,200	20,400	21,600	22,800	24,000	25,200	26,400	27,600	28,800
2.2	14,300	15,400	16,500	17,600	18,700	19,800	20,900	22,000	23,100	24,200	25,300	26,400
2.0	13,000	14,000	15,000	16,000	17,000	18,000	19,000	20,000	21,000	22,000	23,000	24,000
Average value—year 2000												
1.8	**11,700**	12,600	13,500	14,400	15,300	16,200	17,100	18,000	18,900	19,800	20,700	21,600
1.6	10,400	11,200	12,000	12,800	13,600	14,400	15,200	16,000	16,800	17,600	18,400	19,200
1.4	9,100	9,800	10,500	11,200	11,900	12,600	13,300	14,000	14,700	15,400	16,100	16,800
1.2	7,800	8,400	9,000	9,600	10,200	10,800	11,400	12,000	12,600	13,200	13,800	14,400
1.0	6,500	7,000	7,500	8,000	8,500	9,000	9,500	10,000	10,500	11,000	11,500	12,000
Minimum value—year 2000												
0.8	5,200	5,600	6,000	6,400	6,800	7,200	7,600	8,000	8,400	8,800	9,200	9,600
Population (million of units)	**6,500**	7,000	7,500	8,000	8,500	9,000	9,500	10,000	10,500	11,000	11,500	12,000

Note: bold numbers are 2000 year actual figures. In the second column Total energy consumption and Population are shown

3.3 Energy End Users

Final energy consumption can be analyzed by taking the energy demand in each sector into account: industry, transport, residential and nonresidential buildings, agriculture, and non-energy uses. Final consumption is roughly 70 % of the gross world consumption because of losses mainly in electric power production plants and in distribution and in other transformations inside energy industries. For a typical industrialized country the final consumption could be shared among sectors roughly as follows: 30 % industry, 25 % transport, 28 % residential and nonresidential buildings, 2 % agriculture, 5 % non-energy uses, and 10 % energy industry.

The electric energy, in TOE, is roughly one third of the total primary energy consumption.

These balances are shown in Fig. 3.1.

3.3.1 Industry

Industrial energy consumption concerns mainly a few sectors only. Iron and steel, nonferrous metals, nonmetallic minerals, and chemicals account for 40 % of the energy consumed by industry with a tendency to decrease. Four other sectors, i.e., paper and pulp, textiles, food processing, and glass are responsible for

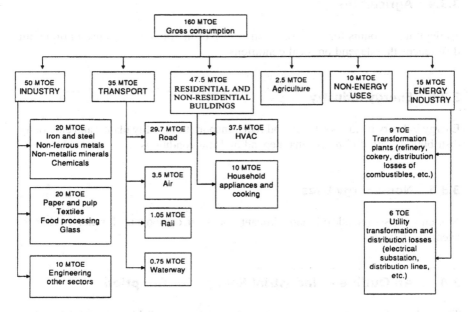

Fig. 3.1 Gross and final consumption in a typical industrialized country with a gross consumption of 160 MTOE and a population of 50 million people

another 40 %. The remaining energy is shared among other manufacturing sectors such as engineering, plastics, electronics, etc.

3.3.2 Transport

More than other sectors road transportation is responsible for energy consumption. Average values for an industrialized country are as follows: 80–85 % for road transport, 10–15 % for air, 3–5 % for rail, 1–3 % for water transport.

Differences will be found depending on local conditions.

3.3.3 Residential and Non-residential Buildings

Residential energy consumption in industrialized countries occurs mainly in heating, ventilation, and air conditioning, which account for roughly 75–80 % if hot water production is included. Household appliances are responsible for 15–20 % of the total consumption and cooking for a further 5 %.

Nonresidential energy consumption (offices, shopping centers, hospitals, public buildings, universities and schools, airports, etc.) occurs mainly in lighting, heating, ventilation, and air conditioning.

3.3.4 Agriculture

Agriculture accounts for not more than 1–2 % of the final gross consumption with differences that depend on local conditions.

3.3.5 Energy Industry

Energy industry includes refinery and fuel transportation, utility plants, and electricity distribution, with differences that depend on local conditions.

3.3.6 Non-energy Uses

Non-energy uses include oils and lubricants, bitumen, components for motor vehicles, chemical industries, etc.

3.4   An Outline of Industrial Energy Consumption

The outline of industrial energy consumption reported in Table 3.2 is based on data from international reports and specific energy audits.

Table 3.2 Specific energy consumption of manufacturing industries

Industrial sector	Heat		Electricity
	t_{oil}/t	MWh/t	MWh/t
Brewing	0.05	0.58	0.10
Brick	0.075	0.87	0.05
Cement	0.08	0.93	0.11
Chocolate	0.20	2.33	2.00
Dairy industry	0.15	1.74	0.5
Engineering	0.30	3.49	2.75
Flour products (pasta, etc.)	0.04	0.47	0.16
Foundry	0.3	3.49	0.9
Ham, sausage	0.1	1.16	0.35
Hot pressing	0.25	2.91	0.75
Ice cream and cake	0.1	1.16	0.75
Milk processing	0.02	0.23	0.10
Nonferrous metal	0.07	0.81	0.30
Paper and pulp	0.15	1.74	0.43
Plastic	0.05	0.58	0.55
Rubber	0.10	1.16	5.00
Textiles (dyeing)	0.75	8.72	0.75
Textiles (spinning)	0.45	5.23	7.5
Wood	0.02	0.23	0.06

Notes: To convert end user heat in t_{oil}/t into MWh/t multiply by (41,860/3,600) (see Sect. 2.3 for this assumption)
Consumption range is ±25 % around the average values shown in the table

Final consumptions as heat (t_{oil} and thermal MWh) and electricity (electrical MWh) are referred to the production (t). Although these values are average ones, they qualify industrial sectors from the energy point of view. Final users' energy costs are not considered as they change according to the use (direct in furnaces and dryers, or to produce steam, or electricity, etc.), the quantity, the geographical area, and the country policies.

These parameters TOE/t have been chosen in order to avoid any relationship between energy consumption and other reference units such as number of employees or company incomes, which are often considered but which may become outdated in a short period, independently of energy consumption.

Utility Plants and Renewable Sources

4

4.1 Introduction

Energy is transported to the site as purchased fuel (oil, gasoil, LPG, natural gas, etc.) or as electricity purchased from utilities. Water, to which attention must be paid when an energy management program is implemented, can be either purchased from utility or pumped from wells.

In addition, energy can be used from the so-called renewable sources, i.e., waste recovery and permanent natural power sources such as sun, wind, geothermal, wave—tidal and ocean thermal energy, and water. The importance of these power sources, which are mainly transformed in electricity, depends on their geographical location, distance from main networks, and the operating period that generally does not exceed 2,500 h/year.

4.2 Utility Plants and Renewable Sources, from Input Energy to End Users

Utility plants, mainly producing electric energy and heat, may be classified as fossil fuel or nuclear plants. In some local conditions, utility plants use other sources, namely hydro—geothermal—solar concentration—wind.

Utility plants generally use steam-condensing turbines, gas turbines, and combined cycles. Heat recovery for district heating is a common way to improve overall efficiency; of course local regulations, climatic conditions, and plant location determine the attractiveness of this possibility.

As a general rule, utility plants' efficiency ranges between 35 and 60 %; if electric distribution losses are considered, an average value of 33–57 % can be assumed for the overall efficiency from primary energy to end users. This means that 0.15–0.24 kg of equivalent oil is required to deliver 1 kWh to the end user.

G. Petrecca, *Energy Conversion and Management: Principles and Applications*,
DOI 10.1007/978-3-319-06560-1_4, © Springer International Publishing Switzerland 2014

Cogeneration plants (see Chap. 9), that is, plants in which either the coincident generation of necessary heat and mechanical or electric power occurs, or which produce power by recovering low-level heat from process, may reach an overall efficiency ranging from 60 to 85 %, depending on the type of cogeneration plant. This is a considerable primary energy saving, generally accompanied by a significant energy cost saving, more or less important depending on the local regulations and tariffs.

The main renewable sources of energy are solar, wind, geothermal, hydraulic energy, and energy from waste. Most of these sources are inadequate to cover the request of a single site, factory, or building, because they yield only low specific power in standard conditions while the current capital and operating costs are rather high. Nevertheless, in spite of these limitations, renewable sources should be considered in the light of local conditions and efforts always made to exploit them.

4.3   Renewable Sources: Solar Energy

The solar energy reaching the earth surface can be used to produce hot water or to produce electric energy by means of photovoltaic cells. Typical values of the rate of the solar energy are 300–1,000 W/m^2 depending on the latitude, time of day, and atmospheric conditions.

4.3.1   Hot Water from Solar Energy

Water can be heated by solar collectors which collect a fraction of the incident sunlight ranging between 90 and 20 % depending on the number and type of glazing and on operating parameters such as rate of insulation, ambient air temperature, and input cooling fluid temperature.

> **With average values of 300–1,000 W/m^2 of solar energy rate, a standard collector may yield an actual specific power ranging from 200 to 600 W/m^2 to produce hot water at 50–60 °C (122–140 °F) with a difference of 30 °C (48 °F) between the average internal collector temperature and ambient temperature.**

Heat losses from a collector are generally defined by a coefficient ($W/m^2 \times K$) which varies from 7 with a single glazed flat plate to 2.5 in the case of a double one with selective surfaces (which have a selective coating such as black chrome to reduce the emission of energy, particularly at high temperatures). The solar energy, when converted into hot water or hot air, must be stored to match the end user demand throughout the day.

Insulated water-storage tanks are always installed; the average capacity is at least 100 L/m^2 of collector (2.4 gallon/ft^2). If an intermediate fluid is used, heat exchangers with high effectiveness must be installed.

With hot-air systems, thermal storage with rock bed can be used, if necessary. Generally, these systems are difficult to justify from the economic and technical points of view for use in factories and in buildings.

4.3.2 Photovoltaic Systems

Electric energy can be produced from solar energy by using photovoltaic silicon cells. As a general indication, the voltage across one cell is 0.5 V and it is independent of the solar energy rate to which the electric current fed to the load is linearly related. In practice, cells are mounted on a module and the modules are arranged in panels or arrays in series, in parallel, or in combined series/parallel.

> **The efficiency, that is the ratio between the electric power generated and the solar energy rate, is not more than roughly 15 % for standard applications. That means that 1 m^2 of cells will produce not more than 75 W with an average solar energy rate of 500 W/m^2.**

The electric energy produced by a series of cells must be stored in an accumulator storage system.

The equivalent operating hours at the peak power range from 1,000 to 2,000 h/years depending on the geographical location.

4.3.3 Solar Concentrated Solar Power Systems

Solar energy is used to heat a fluid at high temperatures by means of concentrator systems like parabolic collectors (concentration factor less than 100), solar tower (concentration factor 300–1,000), or parabolic concentrators (concentration factor 1,000–3,000). The greater the concentration factor is, the higher the temperatures of the heated fluid (water and steam), in most cases from 300 to 600 °C and more.

A classical thermodynamic cycle is then used to produce electricity.

Solar concentration systems are suitable for utility plants, with an electric output power higher than 10–20 MW (the surface of concentrator is 50,000–100,000 m^2) and a total efficiency, the ratio between input solar and output electrical energy, about 30 %. Notice that in case of solar tower and parabolic concentrators heat storage is possible before entering the electric power section of the plant.

4.4 Renewable Sources: Wind Energy

Wind energy has traditionally been exploited for mills at sites where wind velocities are high and quite steady all the year round.

A basic relationship, which relates wind velocity to the power available and to the device surface normal to the wind velocity, is

$$P = (1/2) \times A \times \rho \times V^3$$

where P = power yielded by the wind source (W), A = area normal to the wind velocity (m^2), ρ = air density (1.29 kg/m^3 or 0.08 lb/ft^3 in standard conditions), and V = velocity of airstream (m/s),

For practical applications, the ratio P/A is generally introduced; then

$$(P/A) = (1/2) \times \rho \times V^3 = 0.645 \times V^3 \qquad (W/m^2)$$

$$(P/A) = (1/2) \times \rho \times V^3 = 0.059 \times V^3 \qquad (W/ft^2)$$

The theory of wind devices shows that not more than 59.3 % of the power available in the wind can be converted into mechanical power (Betz limit). Actual values are lower, due to aerodynamic efficiency being generally not higher than about 75 %.

The power coefficient C_p, defined as the ratio between the power actually converted to mechanical power (P_m) and the power available in the wind (P), has an average value of less than 75 % × 59.3 % = 44 %.

Thus

$$(P_m/A) = C_p \times (P/A)$$

Note that the power actually converted into mechanical power by a specific device varies with the cube of wind velocity. Mechanical power may be used to drive equipment such as mills or to produce electric energy by means of direct current (d.c.) or alternating current (a.c.) machines acting as generators. The energy can be delivered to electric networks or stored as d.c. energy in batteries and then converted to a.c. energy by means of a d.c./a.c. static converter.

Wind power converted to mechanical power with a wind velocity of 7–8 m/s is roughly 100–150 W/m^2 (square meter of area normal to wind velocity).

Wind power installations, with a wind velocity of 7–8 m/s, reach up to more than 2,000 kW of mechanical power per single device with a diameter larger than 70–80 m and blades shaped to increase the surface normal to the wind velocity.

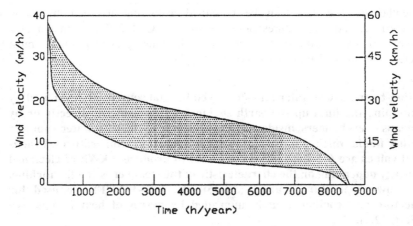

Fig. 4.1 Range of typical variations for annual average wind velocity–duration curves (each wind velocity point on the curve corresponds to the number of hours during which this velocity has been exceeded)

In order to quantify the energy generated by wind devices, speed duration curves for the specific site must be known, that is, wind velocity versus hours per year. Figure 4.1 shows a range of values which can be found in various sites where wind power installations can conveniently be used.

Notice that wind devices generally work in a range of wind speed between cut-in speed (bottom speed) and cutoff speed (upper speed). The wind device works at variable speed; more or less sophisticated controls and power electronic conversion systems are used to improve the exploitation of the available energy. The blade pitch can also be varied.

Of course, the choice among different systems must be based on the capital cost, energy cost saving, and operating costs (see Chap. 19).

4.5 Renewable Sources: Geothermal Energy

Geothermal energy is the natural heat within the earth's crust. Natural-occurring water from aquifer with a temperature range of 50–150 °C (122–302 °F) can be used for building heating, district heating, agriculture, and spas. Water and steam at temperatures higher than 150 °C are required to produce electricity. The fluids are generally re-injected in the ground by injection holes; alternatively, used water may be discarded to the sea or to another body of water.

During the exploitation of geothermal energy there are a few important issues that should be considered, associated with the fact that fluids contain large amounts of diluted salts and gases with potential environmental impacts (atmospheric and surface water pollution). Ground subsidence and noise are other potential major impacts.

Geothermal resources can be classified according to their hydrological characteristics and temperatures: hot water till 250 °C (482 °F), steam at a temperature of 250–300 °C (482–572 °F), hot dry rocks where pressurized water must be injected, and pressurized water in confined spaces.

> **High-temperature water and steam geothermal sources are exploited by pumping the fluid up the earth surface and by using it, directly or by means of exchangers, inside a thermodynamic cycle, with water vapor or other fluids which can work at lower evaporation temperatures. Practical values are 6–8 kg$_{steam}$ (13.2–17.6 lb) to produce 1 kWh of electrical energy depending on the characteristic of the steam crossing the turbine.**
>
> **Low-temperature sources (50–150 °C, 122–302 °F) are used for heating and cooling, usually upgraded by means of heat pumps (see Sect. 12.9).**

4.6 Renewable Sources: Wave, Tidal, and Ocean Thermal Energy

Wave motion and tidal power can be used to produce electricity. Wave energy is related to water height, from hollow to crest, and to the kinetic energy due to the water motion. Tidal energy is related to the exploitation of the oscillatory motion of the water level that occurs periodically and in a predictable way. Ocean thermal energy conversion uses the heat energy stored in the Earth's ocean to produce electricity.

The devices that capture the tidal energy and transform it into electricity are classified according to the operation principle: (1) the potential energy related to the variation of the sea level is captured by means of barrages and hydraulic turbine (see also Kaplan turbine in Sect. 4.7), and (2) the kinetic energy of tidal currents is captured by underwater turbines, like those used in the case of wave energy.

> **The electric power from tidal sources ranges from 0.5 to 1,000 MW, depending on the tidal amplitude and dimension of the water storage. The yearly energy production is strictly related to the time interval between two subsequent tides.**

The devices that capture the energy within the waves and transform it into electricity, called WECs (Wave Energy Converters), are generally classified in four categories, according to the operation principle (oscillating water column, overtopping device, wave-activated bodies), the location (shore, offshore), the power takeoff system (air turbines, linear generators, hydraulic systems, others), and the directional characteristics, i.e., the direction in which the energy is absorbed by the device (point absorber—all directions, terminator—only one direction, attenuator—in parallel with wave direction).

The average annual wave power ranges from 5 to 20 kW/m on the basis of 8,760 h/year with a wave depth from 30 to 60 m depending on the distance from the shore and on the position in the ocean. The power of a single device is generally less than a few hundreds of kW.

The devices that convert ocean thermal energy into electricity can best work when the temperature difference between the warmer top layer and the colder deep ocean water (mainly in tropical oceans) is about 20 °C (36 °F). They must be installed close to the shore and require a large-diameter intake pipe to bring the cold water to the surface. Warm water evaporates at low pressure (1) in specially designed evaporators and then expands in a vapor turbine (2) in a vacuum chamber where the flash-evaporated water vaporizes a low-boiling-point fluid in a closed loop that drives a turbine producing electricity (see also ORC cycle, Sect. 9.2). Deep cold water is used to condensate the vapor in both devices.

Potential heat recovery and production of freshwater are additional advantages of the ocean thermal energy conversion. Typical power ranges from 20 to 30 kW to few MW.

4.7   Renewable Sources: Hydraulic Energy

Hydroelectric power stations are designed to convert the gravitational energy of water into mechanical and then into electric energy. Both large and small power stations can be classified as follows: (1) run-of-river power station when no significant regulating reservoir exists; (2) pondage power station which is run-of-river with small reservoirs able to delay the production of electricity for a short period (hours, days); (3) seasonal power station with a reservoir to regulate the water supply to the turbines; and (4) pumped-storage power station when a reservoir is filled exclusively or partially by pumps.

Large power plants, greater than a few MW, belong exclusively to utilities; small and mini-power stations down to few hundreds of kW usually belong to single factories for their own use.

Three main types of turbine are employed in hydroelectric power stations.

Propeller turbines, such as the Kaplan turbine, are employed for low heads of not more than 40–50 m (130–160 ft) and high flow rates ranging between 2 and 40 m³/s (70–1,400 ft³/s). They provide high rotor velocity for relatively low water through-flow velocities and may attain a good

(continued)

(continued)

efficiency over a wide range of loads by blade pitch variation. Francis turbines are used for higher heads (20–300 m, 66–1,000 ft) and medium flow rates (0.2–20 m³/s, 7–700 ft³/s); these turbines are radial inflow units where water enters the rotor through a set of variable-angle inlet guide vanes and flows radially inward and axially downward, with a pressure drop within the turbine wheel itself.

Pelton turbines, which are impulse units, are used for heads from 100 m (330 ft) up to 300 m (1,000 ft) or higher, with a very low flow rate, generally lower than 1 m³/s (35 ft³/s). All the static head is converted into velocity and all this energy is absorbed in the wheel so that water leaves at a very low velocity.

Since electric power generation at constant frequency requires constant speed, problems arise in the partial-load operating mode when shaft torque is reduced because of a reduction in electric power demand.

Kaplan and Francis turbines can accept partial flow without significant loss of efficiency by varying either the rotor blade or the inlet guide vane angles. Pelton turbines can be regulated by an adjustable nozzle.

The basic relationship giving the power yielded at the turbine shaft by hydraulic energy is as follows:

$P = q \times H \times \rho \times (g/g_c) \times \eta/1,000$ (kW)	$q \times H \times \rho \times (g/g_c) \times 0.0226 \times \eta/1,000$ (kW)
where P = turbine power shaft output (kW)	(kW)
q = volume flow rate (m³/s)	(gpm)
H = total head = $H_{static} - H_{losses}$ (m)	(ft)
ρ = density of water (kg/m³)	(lb/gallon)
η = turbine efficiency	
g = gravitational acceleration 9.81 (m/s²)	32.17 (ft/s²)
g_c = conversion factor = $\frac{1 \text{ kg} \times (\text{m/s}^2)}{N}$	$32.17 \frac{\text{lb} \times (\text{ft/s}^2)}{\text{lbf}}$

On the assumption that $\eta = 1$, the basic relationship shows that with a 100 m head the power related to a flow rate of 1 m³/s equals 981 kW. Thus, 1 kWh of mechanical or electric energy theoretically corresponds to 367 m × m³ (i.e., 367 m head and 1 m³ of falling water) and vice versa for pumps (see Chap. 10).

Figures 4.2 and 4.3 show the ranges of application of the various types of hydraulic turbine units and the effects of load on the efficiency of typical turbines, which is generally higher than 90 %.

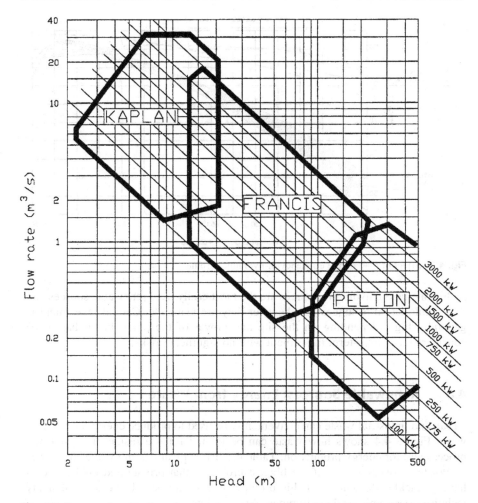

Fig. 4.2 Ranges of application of various types of hydraulic turbine units (100–3,000 kW)

The same relationship can be used in pumping mode by introducing the total head $H = H_{static} + H_{losses}$ and the pump efficiency (see Sect. 10.2).

In choosing small and mini plants one must consider that:

- the useful available inflow may fluctuate during the year or the day;
- usually, such plants are economically viable only if fully automated and free from maintenance;
- the lower the installed capacity, the higher the capital and operating costs per kW installed. Notice that the only operating costs are those for maintenance, which are generally low in terms of kWh cost;

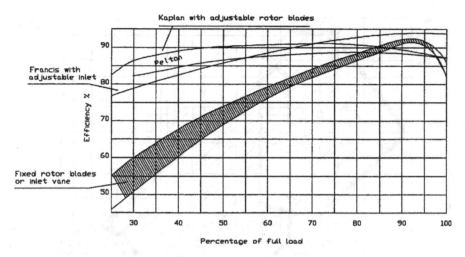

Fig. 4.3 Effect of load on the efficiency of hydraulic turbines

In conclusion, hydraulic sources are the most easily used renewable sources and their exploitation is suggested whenever they are available. Capital investment is generally high, depending on the site and power of the plant, but it has to be considered as a long-term investment with no significant risks.

4.8   Renewable Sources: Energy from Waste

Urban and industrial waste can conveniently be used to produce energy in different ways. Figure 4.4 shows how waste can be treated for use inside or outside the process and for energy production. Table 4.1 lists typical heating values of waste.

Waste recovery is one of the biggest problems that private and public bodies have to tackle. Resolution of this problem is evolving slowly because of many technical, economic, and environmental constraints.

Mass burning facilities where the refuse is burned as received and processed fuel or refuse-derived fuel (RDF) where the solid waste is processed before burning are the two options that technology makes available.

A more detailed analysis of these problems can be found in Chap. 16.

4.9   Energy Storage: Hydro, Mechanical, Electric, Fuel Cell, Thermal Storage

Storage of energy after its transformation from primary energy into various forms is a deeply felt but rarely satisfied need. For both utility plants and end users, the storage of energy should be a way of improving energy management and reducing

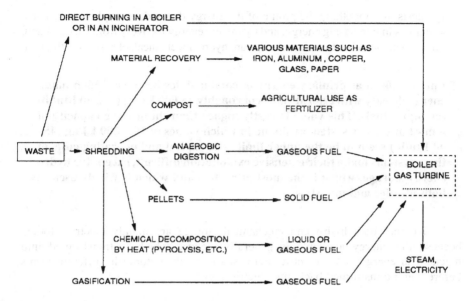

Fig. 4.4 Possible ways of converting waste into energy

Table 4.1 Typical higher heating values of waste		Higher heating value	
	Description of waste	kJ/kg	Btu/lb
	Industrial waste		
	Leather scrap	23,000	9,888
	Cellophane	27,750	11,930
	Waxed paper	27,500	11,823
	Rubber	28,500	12,253
	Tires	41,850	17,992
	Oil, fuel oil residue	41,850	17,992
	Polyethylene	46,000	19,776
	Agricultural		
	Bark	11,000	4,729
	Rice hulls	13,500	5,804
	Corn cobs	19,000	8,169
	Composite		
	Municipal	10,000–15,000	4,299–6,449
	Industrial	15,000–17,500	6,449–7,524
	Agricultural	7,000–14,000	3,009–6,019

energy costs, by smoothing the profile of the energy demand, by exploiting low-rate opportunities in purchasing energy, and by freeing energy recovery from user demand.

The main approaches to energy storage are hydro, mechanical, electric, and thermal.

Energy stored as primary energy in combustibles has a very high mass-energy density with typical values of roughly 45,000 kJ/kg (19,350 Btu/lb) for liquid fuels. This value is vastly higher than the storage capacity of almost any other system available in which values of 100–200 kJ/kg (45–90 Btu/lb) seem to be the upper limit. The capital and operating costs of these systems make their extensive exploitation difficult, except for hydro energy storage, which is the most attractive and which is widely used in both utilities and industries.

Notice that both hydro and mechanical storage are mainly electric storage because the energy available is generally electric which is transformed into mechanical energy at the electric motor shaft and then stored in different forms before being transformed back into electric energy.

4.9.1 Hydro Storage

The principle of hydro storage follows the law of pumps (see Sect. 10.2) and of hydraulic source exploitation by turbines (see Sect. 4.7).

Generally using reversible machines, this system works in the pumping mode when an excess of energy is available (usually electric energy) and in the turbine mode when energy is required back. Water is stored in an upper reservoir and it flows back through the turbine during periods of peak demand.

Remember that energy theoretically associated with 1 m^3 of water and 367 m of head is 1 kWh and that the efficiency of the whole system including energy transformations in pumping mode (electric motor, pump, and flow losses) and in turbine mode (flow losses, turbine, and electric generator losses) is generally 50–60 %. It follows that the energy actually available is halved. A head of more than a hundred meters between the upper and the lower reservoir is usually required for an economic exploitation of the system.

4.9.2 Mechanical Storage

The flywheel is a typical device for the storage of mechanical energy. The quantity of stored energy depends on the shaft speed, the mass, and the radius of gyration of the flywheel (that is, the radius where the total mass is considered to be concentrated).

The energy stored in a flywheel is equal to the kinetic energy, that is,

$$E_{stored} = \frac{1}{2} \times \frac{1}{g_c} \times M \times (R \times \Omega)^2$$

where E_{stored} = energy stored (J),

g_c = conversion factor = $\frac{1\,kg \times (m/s^2)}{N}$,

M = mass of the flywheel (kg),

R = radius of gyration (m). In the case of a disc of uniform density, with uniform thickness and outer radius R_o, $R = 0.7 \times R_o$,

Ω = revolution speed (rad/s); typical values (200–1,000 rad/s).

The energy absorbed or released from a flywheel between two rotational speeds (Ω_1 and Ω_2) is

$$\Delta E = \frac{1}{2} \times \frac{1}{g_c} \times M \times R^2 \times \left(\Omega_1^2 - \Omega_2^2\right)$$

The specific energy referred to the mass of the flywheel depends on the revolution speed, stress-to-density ratio, and geometry of the flywheel. Typical maximum value for high-speed composite-material flywheels is 100 kJ/kg; lower values such as 20–30 kJ/kg are generally reached for low-speed flywheels with isotropic materials and constant-stress disc geometry.

4.9.3 Electric Storage

Electric energy can be stored directly by using electric batteries, among which the lead-acid type is the commonest.

The use of electric batteries for storage depends on many parameters such as charge and discharge cycle life, energy-mass and power-mass ratios, and of course energy-cost ratio.

Table 4.2 lists typical operating parameters of industrial batteries such as lead-acid, nickel/iron, nickel/cadmium, sodium/sulfur, and zinc/bromine.

The specific energy values range between 108 and 114 kJ/kg (30–40 Wh/kg) for lead-acid batteries and 360 kJ/kg (100 Wh/kg) for sodium/sulfur batteries; the specific power values range between 100 W/kg for lead-acid batteries and 400 W/kg for nickel/cadmium batteries; the life cycle does not exceed 400 cycles for lead-acid batteries against 1,000 cycles for nickel/cadmium batteries. The charge–discharge efficiency can reach 60–70 % depending on the operating conditions and on the recharge equipment.

Table 4.2 Operating parameters of various batteries

	Specific energy (Wh/kg)	Specific power (W/kg)	Cycle life to 80 % (cycles)
Lead-acid	40	90	400
Nickel/iron	50	100	>1,000
Nickel/zinc	60	100	250
Zinc/bromine	50	90	200
Nickel/cadmium	50	400	>1,000
Lithium aluminum/iron sulfide	80	70	200
Sodium/sulfur	100	120	250
Iron/air	70	90	150

These values, although higher than those of flywheel storage, are still far from the energy storage capacity of combustibles, thus limiting this kind of storage to particular applications for short periods of demand.

In addition to industrial applications, electric storage affects electric vehicle diffusion. With typical consumption per kilometer and per metric ton of 490 kJ/t × km (0.15 kWh/t × km) if measured at the utility delivering node, a wide use of these vehicles is strictly dependent on the availability of high-performance electric storage systems. With an equal division of the gross weight among payload, vehicle body and traction equipment, and batteries, the urban range is about 70–80 km; this range could be improved by using higher energy-density batteries and by reducing vehicle weight through a proper design of the body and of the traction equipment.

4.9.4 Fuel Cells

Fuel cells are electrochemical devices in which electric energy is produced by combining hydrogen and oxygen and by releasing water vapor into the atmosphere. Basically, the main components of a fuel cell are the anode, the cathode, and the electrolyte (liquid, solid, membrane) between them. To obtain the required voltage a stack of many fuel cells in series is required.

Fuel cells can be classified according to the working temperature (see Table 4.3 where the main operating parameters are reported) as (1) low-temperature cells (AFC, PEM, PEFC, PAFC) and (2) high-temperature cells (MCFC, SOFC).

Hydrogen can be produced by different techniques: natural gas steam reforming, electrolysis by using electric energy, coal gasification, and waste or biomass gasification. Table 4.4 reports the main operating parameters of hydrogen production and of electricity production by using hydrogen-fed fuel cells.

Table 4.3 Operating parameters of fuel cells

	Low temperature			High temperature	
	AFC	PEM, PEFC	PAFC	MCFC	SOFC
Electrolyte	Potassium hydroxide	Proton exchange, polymeric membrane	Phosphoric acid	Lithium carbonate, potassium carbonate	Zirconium oxide
Operating temperature (°C)	60–120	70–100	160–220	600–650	800–1,000
Operating temperature (°F)	140–248	158–212	320–428	1,112–1,202	1,472–1,832
Catalyst	Platinum, palladium, nickel	Platinum	Platinum	Nickel	Not necessary
Materials for construction	Plastics, graphite	Metals, graphite	Graphite compounds	Nickel, inox steel	Metals, ceramic materials
Oxidant	O_2	Air	Air	Air	Air
Electric efficiency (%)	60	40–60	50	45–55	45–60
Power density (mW/cm^2)	300–500	300–900	150–300	150	150–270
Power range (kW)	5–80	5–250	<11,000	<2,000	100
Starting time	Minutes	Minutes	1–4 h	5–10 h	5–10 h
Applications	Space, transportation	Small cogeneration plants, transportation	Cogeneration plant	Cogeneration, industrial plants	Cogeneration, industrial plants
Advantages	High power density	High power density, low corrosion, reduced starting time	High-efficiency cogeneration	High-efficiency cogeneration, high-temperature recoverable heat, internal reforming	High-efficiency cogeneration, high-temperature recoverable heat, internal reforming, no catalyst
Disadvantages	Low resistance to CO, high quality of hydrogen	Low resistance to CO, presence of water	Low resistance to CO	Short material life, CO_2 recirculating, long starting time	High temperature, long starting time

Low temperature: *AFC* alkaline fuel cell, *PEM* proton exchange membrane fuel cell, *PEFC* polymeric electrolyte membrane fuel cell, *PAFC* phosphoric acid fuel cell, High temperature: *MCFC* fuse carbonate fuel cell, *SOFC* solid oxide fuel cell

Table 4.4 Operating parameters of hydrogen production and electricity production by fuel cells

		Natural gas steam reforming	Water electrolysis	Coal gasification	Waste gasification	Biomass gasification
Hydrogen production	Nm³	1	1	1	1	1
	kJ	10,700	10,700	10,700	10,700	10,700
Input natural gas	Sm³	0.46				
	kJ	16,307				
Input electric energy	kWh		5.8			
	kJ		20,880			
Input coal	kg			0.53		
	kJ			15,225		
Input waste	kg				2	
	kJ				26,000	
Input biomass	kg					1.4
	kJ					21,000
Hydrogen production energy efficiency (a)	%	66 %	51 %	70 %	41 %	51 %
CO₂ local emission	kg	0.9	0	1.47	1.16	0.44
Electricity production						
Fuel cell average efficiency (b)	%	50 %	50 %	50 %	50 %	50 %
Total efficiency (a) × (b)	%	33 %	26 %	35 %	21 %	25 %
Reference values						
Natural gas LHV	*kJ/Sm³*	*35,450*				
Coal LHV	*kJ/kg*			*29,000*		
Waste LHV	*kJ/kg*				*13,000*	
Biomass LHV	*kJ/kg*					*15,000*

> **Fuel cells, operating on non-petroleum fuels such as hydrogen, might provide an alternative energy source for electric traction and for other applications with a total efficiency, that is, combined hydrogen and electricity production efficiency, of roughly 25–40 %.**

4.9.5 Heat and Cold Storage

Heat and cold storage can be classified as sensible and latent energy storage.

The first group includes systems where the storage is accomplished by increasing or decreasing the temperature of the material (water, organic liquid, solid); the storage energy density depends on the temperature change and specific heat of the material. The evolution of the temperature transient to reach the desired temperature is roughly exponential and the time constant, that is, the time needed to reach

63.2 % of the final temperature in a simplified model, depends on the product of the heat capacity of the mass involved multiplied by the thermal resistance (see Sect. 8.3) of the system.

The energy stored as sensible heat in the material at steady state, when the temperature transient has been completed, is

$$E_{stored} = c \times M \times (t_2 - t_1)$$

where E_{stored} = energy stored (kJ, Btu), c = specific heat of the material to be heated (kJ/kg $\times$ K, Btu/lb $\times$ °F), M = mass to be heated (kg, lb), $c \times M$ = heat capacity of the mass to be heated (kJ/K, Btu/°F), and t_1, t_2 = temperatures of the mass to be heated or cooled (K, °C, °F).

> If mineral oil is used for heat storage, the storage energy density associated with a temperature increase of 50 K (50 °C; 90 °F) is roughly 100 kJ/kg; the storage energy density with water for the same temperature increase is roughly 200 kJ/kg.
>
> If water is used for cold storage, a temperature drop of 4–5 K (7.2–9 °F) is usually obtained, so that the storage energy is 20 kJ/kg.

The second group includes systems where the energy is mostly stored in the form of the latent heat due to a phase change, such as melting a solid (the opposite occurs in cold storage) or vaporizing a liquid. In releasing energy, liquids solidify (or solids liquefy) and vapors condense. The storage energy density per kg, which derives mainly from the latent heat, is greater than in sensible energy storage systems. An additional advantage of this group is that it works at constant temperature during the phase change.

In addition to proper transition temperatures and high latent heat, materials for storage purposes must possess other physical and chemical properties such as thermal conductivity, stability, and non-toxicity in the operating conditions. Eutectic salts such as NaF-FeF_2 and $ZnCl_2$ with a density of roughly 2,000 kg/m^3 may reach energy densities of 400–1,500 MJ/m^3 or 200–750 kJ/kg (10,700–40,300 Btu/ft^3, 90–340 Btu/lb) with an overall efficiency of not more than 50 %. Capital and operating costs are quite high and do not allow a wide exploitation of this system in industry.

> If ice is used for cold storage, an average storage capacity of 350 kJ/kg can be reached, of which 335 kJ/kg comes from the latent heat and 15 kJ/kg from the sensible heat. The electric energy required for producing the ice depends on the COP of the refrigerating system; if COP equals 3 (which can be considered as a typical value), the consumption is 0.032 kWh/kg of ice produced in the storage system.

4.10   Applicability in Factories and in Buildings

A site's demand of both electricity and thermal energy is often satisfied by means of utility plants and related distribution lines. Generally, no alternative source of thermal energy is available. On the other hand, electric energy can be produced on site from either traditional fuels or refuse-derived fuels.

Cogeneration and trigeneration plants and waste-recovery plants seem to be the most attractive means of reducing primary energy consumption and the cost of the energy supply, but the capital and operating costs must be investigated together with technical aspects such as the quality, reliability, and flexibility of the energy production.

Renewable sources can provide significant support only in particular conditions. Nevertheless, energy management programs must always verify these possibilities.

Electrical Substations

5

5.1 Introduction

Electric energy is commonly delivered to industrial users by three-phase networks at medium or high voltage. The values of medium and high voltage are defined by international standards; broadly, medium voltages range from 5 to 25 kV and high voltages range from 50 to 500 kV. Higher and lower values can be found according to local conditions.

Individual industrial users have to reduce the levels indicated above to lower voltages inside factories. Typical low voltage values range from 260 to 380 V.

Transformers at site boundaries and related electrical equipment make up the electrical substations, through which all the purchased electric energy flows, and they are responsible for transformation losses, as are boiler plants with thermal energy (see Chap. 6).

Although the transformer's efficiency is quite high (generally not less than 98 %) it is worth reducing these losses as much as possible, because their quantity is related to the total electric energy consumed in the site.

5.2 Basic Principles of Transformer Losses and Efficiency

Transformer losses may be no-load or load losses:
- No-load losses are almost exclusively iron losses. They are a squared function of voltage and occur whenever the transformer is connected to a voltage source, independently of the load value;
- Load losses are a squared function of current (or of apparent power load). They are mainly copper losses and occur only when a load is connected to the transformer, through which current is flowing.

G. Petrecca, *Energy Conversion and Management: Principles and Applications*, DOI 10.1007/978-3-319-06560-1_5, © Springer International Publishing Switzerland 2014

The electrical machine theory shows that the maximum efficiency is reached at a load level at which no-load and load losses are equal:

$$\text{Efficiency} = \frac{\text{Total output power}}{\text{Total input power}} = \frac{\text{Total output power}}{(\text{Total output power} + \text{Total losses})}$$

There follows:

$$\text{Efficiency} = \frac{x \times P_n}{(x \times P_n + P_0 + x^2 \times P_{cn})}$$

where x = load factor = ratio between total actual output and rated output power, P_0 = no-load losses (kW), P_{cn} = load losses at rated power (kW), Total losses = $P_0 + P_{cn} \times x^2$ (kW), $r = P_{cn}/P_0$, P_n = rated output power (kW at fixed power factor; $P_n = A_n \times \cos\varphi_n$), Total actual output = $x \times P_n$ (kW).

Maximum transformer efficiency is reached if:

$$x^2 \times P_{cn} = P_0 \qquad x = \sqrt{\frac{P_0}{P_c}} = \sqrt{\frac{1}{r}}$$

Then, the load value at which maximum efficiency can be reached is a function of the ratio r between load losses at rated power and no-load losses, which depends on the design parameters of the transformer. Let us consider a group of transformers with the same rated power and total losses, but with different values of ratio r; there follows that the maximum efficiency for each transformer corresponds to a specific load factor (x), that is, to an operating point where load losses ($x^2 \times P_{cn}$) equal no-load losses (P_0).

This means that maximum efficiency is reached if, as r increases, the load factor x decreases. Figure 5.1 shows transformer efficiency versus load factor for different values of ratio r for such a group of transformers. The dotted line represents maximum efficiency operating points. Notice that maximum efficiency at rated power ($x = 1$) is reached only if the transformer ratio r equals 1.

If r is higher than 1, which is a typical design criterion, the maximum efficiency can be reached only when the transformer is partially loaded.

Typical values for industrial applications are: r equal to 5–7; maximum efficiency equal to 98.5–99 %; load factor equal to 35–55 %. This means that typical industrial transformers must be loaded well below nameplate values to obtain good efficiency and oversized machines must be installed.

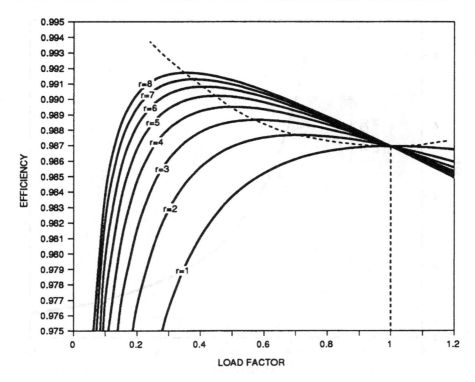

Fig. 5.1 Transformer efficiency versus load factor x for a group of transformers with the same rated power and rated total losses, but with different values of r ratio

As a general energy-management rule it is not advisable to allow transformers to operate constantly at full load.

Figure 5.2 shows the ratio r versus load factor when the maximum efficiency condition is imposed. Typical values for load factor x (35–55 %) and for ratio r (4–7) are also indicated.

5.3   How to Choose the Most Suitable Transformer

Transformer selection is based on many parameters such as capital cost, operating and maintenance costs, energy losses costs, predictable overload conditions, network short-circuit duties, large-motor starting, etc.

Energy losses cost is one of the evaluation parameters although not the most crucial. However, if a correct choice of the nameplate power takes also energy saving into account, worthwhile economies should be obtained. First, it is necessary to assume as reference the average load during a prefixed period, i.e., the load to which the same losses correspond in the real or predicted cyclic conditions of operation. For a low average load, a transformer with a higher r ratio value should

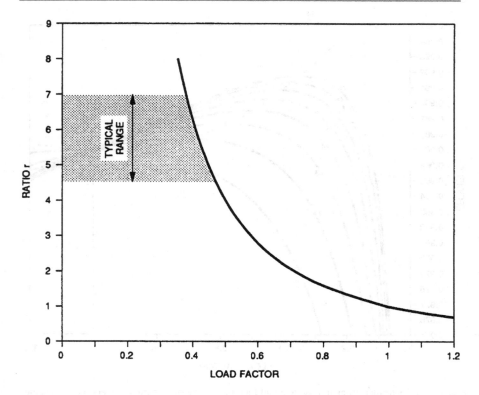

Fig. 5.2 Ratio r versus load factor x if the maximum transformer efficiency condition is imposed

be selected. Finally, provided that the other selection parameters are satisfied, the nameplate power needed for maximum efficiency can be determined, with due consideration to the load average power value. Table 5.1 lists standard transformer parameters. Notice that standard type transformers with reduced losses are to be preferred; only for very large machines or for special applications should non-standard machines be considered.

Table 5.2 and Fig. 5.3 show the influence of no-load and load losses when the load is changing and the value of r ratio is defined. The point of balance between losses corresponds to the maximum efficiency.

Economic evaluations for transformer selection based on energy saving can be made as explained in Chap. 20. As a general rule, notice that in normal conditions the annual losses cost is equal to roughly 10–30 % of the capital cost of the transformer itself. It is worth remembering that in sizing transformers attention must be paid to correlated costs, such as control and protection switchgear, voltage level, cables, foundations, space occupied, etc., which are approximately equal to the cost of the transformer itself, so that the incidence of annual losses is cut in half. Oversizing transformers may thus be more expensive with respect to switchgear

Table 5.1 Standard parameters for oil-insulated transformer MT/BT (normal and reduced losses)

Rated power A_n (kVA)	No-load losses P_0 (W)	Load losses P_{cn} (W)	$r = P_{cn}/P_0$	No-load current (%)	V_{cc} (%)
Normal losses					
50	190	1,100	5.8	2.9	4
100	320	1,750	5.5	2.5	4
160	460	2,350	5.1	2.3	4
250	650	3,250	5.0	2.1	4
400	930	4,600	4.9	1.9	4
600	1,300	6,500	5.0	1.8	4
1,000	1,700	10,500	6.2	1.5	6
1,600	2,600	17,000	6.5	1.3	6
2,000	3,200	22,000	6.9	1.2	6
2,500	3,800	26,500	7.0	1.1	6
Reduced losses					
50	150	850	5.7	1.9	4
100	250	1,400	5.6	1.5	4
160	360	1,850	5.1	1.3	4
250	520	2,600	5.0	1.1	4
400	740	3,650	4.9	0.9	4
630	1,040	5,200	5.0	0.8	4
1,000	1,300	9,000	6.9	0.7	6
1,600	2,000	13,000	6.5	0.5	6
2,000	2,400	16,000	6.7	0.5	6
2,500	2,900	21,000	7.2	0.5	6

Table 5.2 No-load and load losses for a normal-losses transformer at different load factor values (load factor x % ranges from 120 to 10)

Load factor (%)	No-load losses P_0 (W)	Load losses (W)	Efficiency (%)
120	1,700	15,120	98.47 %
110	1,700	12,705	98.57 %
100	1,700	10,500	98.66 %
90	1,700	8,505	98.76 %
SO	1,700	6,720	98.84 %
70	1,700	5,145	98.93 %
60	1,700	3,780	99.00 %
50	1,700	2,625	99.05 %
40	1,700	1,680	99.07 %
30	1,700	945	99.03 %
20	1,700	420	98.84 %
10	1,700	105	98.03 %

Rated power 1,000 kVA; power factor equal to 0.9; $r = 10,500/1,700 = 6.2$

and cables short-circuit rating than with respect to the capital cost of the transformer itself. Energy optimization reduces the annual losses cost, but does not substantially change its incidence on the transformer and related equipment capital cost, which still remains quite considerable.

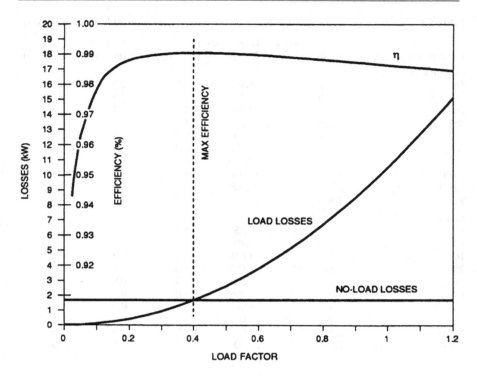

Fig. 5.3 Operating parameters for a standard 1,000 kVA transformer, as specified in Table 5.2

Roughly speaking, a 10 % reduction of losses, corresponding to an increase of 1 % in transformer efficiency, should justify more than 10–15 % of additional capital costs.

5.4   Practical Examples

An example of choice based on energy optimization is given in what follows.

Example 1 Sizing of transformers based on maximum efficiency operation all the year round.

Input data

Load average power	400 kW
Load average power factor	0.85 leading
Operating hours at full load	3,200 h/year
Operating hours at no-load	5,560 h/year
Oil-insulated transformers with technical characteristics as detailed in Table 5.1	
The load is supplied through only one transformer	

Table 5.3 Efficiency of different normal-losses transformers at the same average load power

Transformer rated power A_n (kVA)	Ratio r	Load factor x (%)	No-load losses P_0 (W)	Load losses (W)	Efficiency (%)
400	4.9	117.65 %	930	6,367	98.22 %
600	5	78.43 %	1,300	3,998	98.69 %
1,000	6.2	47.06 %	1,700	2,325	99 %
1,600	6.5	29.41 %	2,600	1,470	98.99 %

Average load power 400 kW; power factor equal to 0.85

Questions
(1) Find the transformer rated power A_n (kVA) that gives the maximum efficiency at full-load operation.
(2) Calculate the efficiency that corresponds to differently rated power transformers with the same 400 kW load.
(3) Calculate the losses that correspond to different transformers at 400 kW load operation and the energy saved, if the maximum efficiency operation is chosen instead of the minimum efficiency operation transformer.

Answers
(1) and (2) By reference to Table 5.1 it is possible to calculate the efficiency that corresponds to 400 kW load power at power factor 0.85 for differently rated power transformers:

$$x = \text{Load factor} = \frac{400 \text{ kW}}{\text{kVA Rated power} \times 0.85}$$

$$\text{Efficiency} = \frac{x \times \text{kVA Rated power} \times 0.85}{x \times \text{kVA Rated power} \times 0.85 + P_0 + P_{cn} \times x^2}$$

Table 5.3 gives the results of the calculations.

Notice that maximum efficiency corresponds to a transformer with a rated power equal to 1,000 kVA.

The corresponding load factor is equal to 47.06 % and the maximum efficiency is equal to 99 %. The minimum efficiency, equal to 98.22 %, corresponds to a transformer with a 400 kVA rated power (rated powers less than this value are not considered, to avoid any thermal problem).

(3) Table 5.4 shows annual energy losses (kWh) for different transformers at 400 kW load operation (see Table 5.3 for values of P_0 and P_{cn}).

Example 2 For a given rated power, choose between a reduced-losses transformer and a normal-losses transformer.

Calculate the annual energy losses of the two transformers and the annual energy saving.

Table 5.4 Total annual losses of different transformers (as in Table 5.3)

Transformer rated power (kVA)	Annual losses at no-load operation (kWh/year)	Annual losses at average load power (kWh/year)	Total annual losses (kWh/year)	Annual energy saving (kWh/year)
400	5,171	23,350	28,521	
600	7,228	16,954	24,182	
1,000	9,452	12,880	22,332	
1,600	144,566	13,024	27,480	
(a)	(b) $= P_0 \times 5{,}560$, see Table 5.3	(c) $= (P_0 + \text{load losses}) \times 3{,}200$, see Table 5.3	(d) $= (b) + (c)$	
case (400)–case (1,000)				6,189

No-load hours 5,560 h/year; load hours 3,200 h/year
Average load power (kW) 400; power factor 0.85

Table 5.5 Comparison between total annual losses of two 1,000 kVA transformers (a reduced-losses and a normal-losses transformer)

Type of operation	Working hours (h/year)	Load factor x (%)	Annual no-load losses (kWh/year)	Annual load losses (kWh/year)	Total annual losses (kWh/year)
Normal-losses transformer ($P_0 = 1{,}700$ W, Table 5.3); ($P_{cn} = 10{,}500$ W, Table 5.3)					
No-load	5,560	0.00 %	9,452	0	9,452
Load	3,200	47.06 %	5,440	7,441	12,880
Total (a)	8,760				22,332
Reduced-losses transformer ($P_0 = 1{,}300$ W, Table 5.3); ($P_{cn} = 9{,}500$ W, Table 5.3)					
No-load	5,560	0.00 %	7,228	0	7,228
Load	3,200	47.06 %	4,160	6,378	10,538
Total (b)	8,760				17,766
Energy saving (a) – (b) kWh/year					4,566

Input data

Load average power	400 kW
Load average power factor	0.85 leading
Transformer rated power	1,000 kVA
Operating hours at full load	3,200 h/year
Operating hours at no-load	5,560 h/year

Reduced-losses and normal-losses transformers with technical characteristics
as detailed in Table 5.1

The load is supplied through only one transformer

Table 5.5 shows the results of the calculations.
The annual energy saving is 4,566 kWh.
The economic evaluation is given in Table 19.4.

Boiler Plants

6

6.1 Introduction

Combustion in boiler plants transfers the energy entering the site as fuel (solid, liquid, natural gas) to intermediate fluids such as steam, oil, air, water, and others. These have the task of carrying energy to the end users at the lowest possible cost and at maximum efficiency.

Two boiler classes, generally called unfired boilers because they do not require fuel consumption, that is, heat recovery or electric/electrode boilers, can provide an alternative to fuel-fed boilers.

If natural gas is used, it could be necessary to install a station where pressure is reduced upstream of the boiler plant, depending on local gas pipeline conditions and burner requirements. Losses due to this transformation are generally low.

End users such as furnaces, ovens, dryers, kilns, etc. generally burn fuel directly without any upstream transformation.

> **Industrial-boiler plant efficiency generally ranges from 80 % to 90 %; this means that not less than 10–20 % of the total fuel purchased is lost in the boundary zones of the site before distribution to the end users. This is a considerable quantity and makes the boiler plant one of the most dissipative facilities; therefore attention must be paid to this plant with regard to both capital investment and maintenance efficiency, in order to minimize losses.**

One of the basic tasks of energy management is undoubtedly to optimize the operation of facilities which are responsible for the greatest part of the energy losses between site boundaries and end users.

Boiler plants are one of the most important facilities because of the great quantity of energy generally flowing through them. Any capital investments in boilers and related equipment (such as combustion control, water treatment, air

G. Petrecca, *Energy Conversion and Management: Principles and Applications*,
DOI 10.1007/978-3-319-06560-1_6, © Springer International Publishing Switzerland 2014

economizer, etc.) must be carefully studied with the aid, if necessary, of detailed engineering analyses, in order to evaluate the return on investments and the benefits to be obtained from both energy-saving and air-quality improvements.

6.2   Basic Principles of Combustion

Combustion is an exothermic reaction of carbon and hydrogen with oxygen to produce carbon dioxide (CO_2) and water vapor (H_2O).

Combustion starts when the ignition temperature is reached (see Table 6.1) and ends when one element of the reaction is exhausted. This means that incomplete combustion will occur if the quantity of oxygen available in the combustion zone is insufficient for the reaction of all fuel molecules. In this case, the combustion will produce effluents that still contain combustibles, and the heat delivered will be less than expected. Secondary combustion products such as NO_x, SO_x, CO, and solid particles are also released. Their influence on air pollution must be carefully considered.

The main reactions for each element are outlined below:
• Carbon
  Complete reaction

$$C + O_2 = CO_2 + 34,032 \frac{kJ}{kg\ Carbon}$$

If this reaction is not completed

$$C + \frac{1}{2}O_2 = CO + 10,465 \frac{kJ}{kg\ Carbon}$$

$$CO + \frac{1}{2}O_2 = CO_2 + 10,172 \frac{kJ}{kg\ CO}$$

Table 6.1 Ignition temperatures of different combustibles

Combustible	Ignition temperature		
	K	°C	°F
Dry wood	523	250	482
Coal	823	550	1,022
Oil	603	330	626
Gasoline	753	480	896
Natural gas	918	645	1,193
Carbon monoxide	883	610	1,130

- Hydrogen

$$H_2 + \frac{1}{2}O_2 = H_2O \text{ liquid} + 141,486.8 \frac{kJ}{\text{kg Hydrogen}}$$

$$H_2 + \frac{1}{2}O_2 = H_2O \text{ vapor} + [141,486.8 - (9 \times 2,511.6)] \frac{kJ}{\text{kg Hydrogen}}$$

$$= H_2O \text{ vapor} + 118,882.4 \frac{kJ}{\text{kg Hydrogen}}$$

(combustion of 1 kg hydrogen produces 9 kg liquid water which can be vaporized by roughly 2,511.6 kJ/kg)

- Sulfur

$$S + O_2 = SO_2 + 10,883.6 \frac{kJ}{\text{kg Sulfur}}$$

> **Most of the oxygen used for combustion comes from atmospheric air: in volume, 21 % of oxygen (O_2), 78 % of nitrogen, and traces of argon and carbon dioxide can be assumed as typical values. Notice that because of the nitrogen in the air, NO_x emissions always occur, depending mainly on flame temperatures.**

Emission control in order to minimize the formation of NO_x, as well as that of CO and SO_2, is a task of boiler-plant maintenance that must be performed with the same care as energy saving. Quite often the two problems are closely related and can be solved together with considerable economies if they are tackled at the same time.

As shown in the reactions described above, water always forms through the combination of hydrogen and oxygen. The water may be in liquid or vapor state, so two different heating values (gross and net values, otherwise named higher and lower heating values) are considered (see Sect. 2.2). In industrial applications, effluents flowing from the boiler are still at a temperature of more than 373.15 K (100 °C; 212 °F), so that the water is always in vapor state.

The theoretical amount of air needed for complete combustion can be calculated as follows:

- The mass M_o (kg) of oxygen which enhances the complete combustion of 1 kg of solid or liquid combustibles is determined by their composition (mainly C%; H_2%; S%; O_2%). The composition of different combustibles is shown in Table 6.2;
- The related oxygen volume V_o is equal to M_o multiplied by the oxygen specific volume in standard conditions, that is:

Table 6.2 Solid, liquid, and gaseous combustibles and combustion parameters

Fuel	Density	Lower Heating Value	Oxygen for combustion	Air combustion volume			Exhaust at		
				Theoretical value	Practical excess air	Practical value	273 / 0 / 32	493 / 220 / 428	523 K / 250 °C / 482 °F
	kg/m³	kJ/kg	Sm³	Sm³	%	Sm³	Sm³	m³	m³
Solid									
Pitch lignite	650	18,000	1.47	7	100.00	14	15	27.1	28.8
Standard coal	700	29,300	1.26	6	80.00	11	12	21.7	22.9
Charcoal	750	31,400	1.89	9	80.00	16	17	30.7	32.6
Gas coke	400	26,800	1.68	8	80.00	15	16	28.9	30.6
(All values are referred to 1 kg of combustible with average values of C, H₂, S)									
Liquid	kg/m³	kJ/kg	Sm³	Sm³	%	Sm³	Sm³	m³	m³
Refined kerosene	800	43,100	2.52	12	20.00	14.5	15	27.0	29
Gasoil	825	42,700	2.478	11.8	30.00	15.3	16.2	29.3	31
Fuel oil No. 2	860	39,000	2.352	10.3	40.00	14.4	15	27	29
Fuel oil No. 6	1,000	43,000	2.352	11.2	40.00	15.7	16.6	30.0	31.8
(All values are referred to 1 kg of combustible with average values of C, H₂, S)									
Gaseous	kg/1,000 Sm³	kJ/Sm³	Sm³	Sm³	%	Sm³	Sm³	m³	m³
Natural gas	750	34,325	1.995	9.5	20.00	11.4	12.2	22.03	23.37
(All values are referred to 1 Sm³ of natural gas with average values of C, H₂)									

$$V_o = M_o \times \frac{22.4}{\text{oxygen molecular weight}} = M_o \times \frac{22.4}{32} \left(\frac{\text{Sm}^3}{\text{kg comb}}\right)$$

where

22.4 m^3/kmol is the standard volume of 1 kmol of any gas at 273.15 K (0 °C; 32 °F) and 0.1 MPa (14.5 psi)

32 kg/kmol is the oxygen molecular weight;

- The theoretical amount of air needed for the complete combustion of 1 kg of a solid or liquid combustible, including the standard proportion of oxygen (21 %) is:

$$V_{\text{th}} = \frac{100}{21} \times M_o \times \frac{22.4}{32} \left(\frac{\text{Sm}^3}{\text{kg comb}}\right)$$

The same procedure can be applied to gaseous combustibles.

Notice that, practically, the amount of air needed to ensure complete combustion is higher than the theoretical one. If the real value is V, the ratio between $(V - V_{\text{th}})$ and V_{th} is called excess air. It is generally expressed as a percentage. Practical values range from 10 % to 100 % depending on the type of combustible (gaseous, liquid, solid).

Practical values of air combustion are: 10.5–11.5 Sm^3 for the combustion of 1 Sm^3 natural gas; 12–15 Sm^3 for the combustion of 1 kg oil.

6.3 Combustion Efficiency

It is customary to express combustion efficiency as the ratio of the heat delivered by the combustion reactions to the fuel input.

Theoretical figures of heat delivered have already been expressed as heating values; actual values may be lower because of oxygen lack or excess in the combustion zone. To ensure complete combustion, fuels are always burned in excess of air and then of oxygen.

Excess air causes excess of flue gas from the combustion zone, so that the real is always lower than the theoretical combustion efficiency value, heat being spent on excess air. In spite of excess air, flue gas always contains some unburned combustibles and so efficiency is further reduced.

Another loss is due to the heat spent on evaporating and superheating the moisture in the fuel that enters the combustion zone and on superheating the moisture in the combustion air. As shown in Sect. 6.2, vapor also comes from burning hydrogen in the fuel, and the heat spent on evaporating and superheating this makes up an additional loss.

Table 6.2 shows typical values of parameters qualifying the combustion of different fuels (heating values, O_2, excess air, exhausts).

In conclusion, combustion efficiency can be improved by keeping the excess air as well as the unburned combustible content as low as possible. That seems to be contradictory, but in this way valuable results can be obtained by optimizing the design, operation, and maintenance of burners and related auxiliaries, as well as that of combustion control equipment.

Combustion losses are generally included among boiler losses, as discussed in Sect. 6.7. Notice that the main contribution to these losses is the heat going up the stack, which is closely related to both the excess air required by combustion and the temperature of the fluids required by the end users.

6.4   Fundamentals of Steam Generation

If boilers are used to produce steam, which is the most frequent case, the heat generated by combustion must be transferred to the water to allow it to reach boiling point and higher temperatures.

There are successive phases:
- Water heating.
  The heat required to bring 1 kg of water from the reference temperature 273.15 K (0 °C; 32 °F) to boiling point is the enthalpy (h_i) or heat content of the liquid. The enthalpy increases with the temperature of the liquid and then with its pressure.

$$h_i = c \times (t - t_R)$$

where c = specific heat of water = $4.186 \frac{kJ}{kg \times K} = 1 \frac{Btu}{lb \times °F}$, t = boiling point temperature (373.15 K; 100 °C; 212 °F), t_R = reference temperature (273.15 K; 0 °C; 32 °F).

At atmospheric pressure at sea level:

$$h_i = 418.6 \frac{kJ}{kg} = 180 \frac{Btu}{lb}$$

If the heating starts at a temperature t_0 above 273.15 K (0 °C; 32 °F), the heat needed to reach boiling point is reduced as follows:

$$h_i = c \times (t - t_0)$$

Liquid water can also be heated to over 373.15 K if the pressure inside the water circuit is kept higher than that of the atmosphere.
- Vaporization.
  When water boils and steam starts to be produced, they are both at the same temperature, that is, the saturation temperature which has only one value for each boiling pressure. During this phase, the temperature remains constant and the heat converts water into steam. This heat is the evaporation enthalpy (h_v), which decreases as temperature and pressure increase.

Thus:

$$\text{enthalpy of saturated steam} = h_i + h_v$$

If saturation occurs at atmospheric pressure, the total amount of heat added to convert water at 273.15 K (0 °C; 32 °F) to 100 % steam is equal to:

$$\text{enthalpy of saturated steam} = 418.6 + 2,252 = 2,670.6\,\frac{\text{kJ}}{\text{kg}} = 1,149\,\frac{\text{Btu}}{\text{lb}}$$

When the mixture of liquid and vapor is balanced, the liquid phase is a saturated liquid and the vapor phase is a saturated vapor.

This mixture can be classified by the quality index x which is the ratio between the mass of vapor present in the mixture and the total mass of the mixture itself.

The value of the quality index x ranges from zero to unity: at saturated liquid state $x = 0$, and at saturated vapor state $x = 1$. The quality index is frequently expressed as a percentage.

For saturated steam, typical values of enthalpy range from 2,670 to 2,770 kJ/kg (from 1,148 to 1,191 Btu/lb).

- Superheating.
  When the water is converted into 100 % steam, the temperature will rise if heat is added. The enthalpy of the steam increases by the amount of added heat, which can be calculated by multiplying the specific heat of the steam (average value at typical operating pressures is 2.1 kJ/kg × K, see Table 2.5) by the temperature increase above the boiling point.

  For a given pressure the temperature depends on the heat added which is called the enthalpy of superheating (h_s). At a given pressure, any temperature above the saturation value can be reached.

  Thus:

$$\text{enthalpy of superheated steam} = h_i + h_v + h_s$$

- Condensation.
  During this phase, which is the opposite of the superheating and vaporization phases, vapor is transformed into liquid water by the extraction of heat. At a given pressure, the superheated steam temperature decreases until the saturation temperature is reached; then, liquid water and steam remain at constant temperature until 100 % liquid water is produced. The temperature depends on the pressure.

  Notice that the heat to extract is equal to the enthalpy of the evaporation phase at the same pressure and temperature.

  Figures 6.1 and 6.2 and Tables 6.3 and 6.4 show the steam properties which are generally obtainable from a Mollier chart.

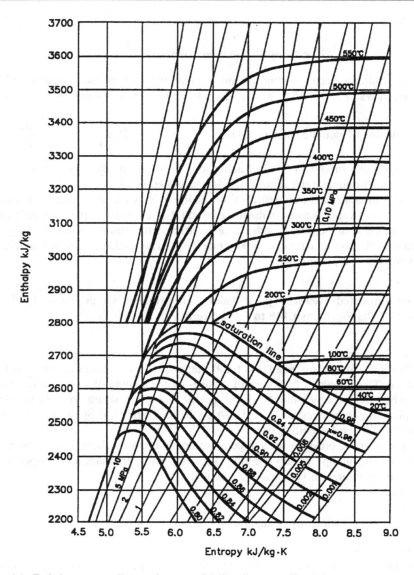

Fig. 6.1 Enthalpy entropy diagram for water (Mollier diagram—SI units)

6.5   Industrial Boilers

Boilers can be regarded as containers into which water or other fluids are fed and there heated continuously by the heat released during combustion. More frequently water evaporates continuously into steam and the state of the fluid is then changed.

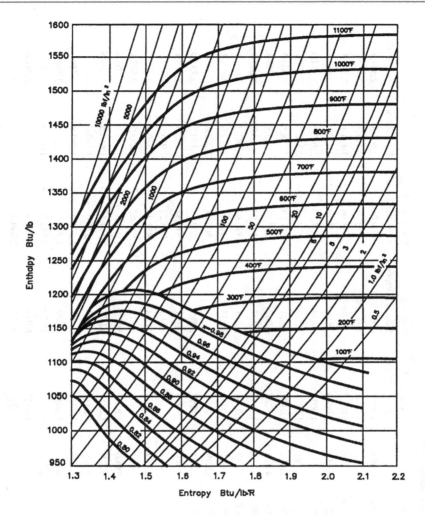

Fig. 6.2 Enthalpy entropy diagram for water (Mollier diagram—English units)

Industrial boilers can be classified as either watertube or firetube depending on the position of the hot combustion gases in relation to the fluid being heated (water, oil). Oil can be used as the heating medium both to carry thermal energy directly to the end users and to produce steam by means of a heat exchanger acting as an evaporator inside or outside the boiler plant.

The so-called unfired boilers such as heat recovery boilers and electric or electrode boilers, can also be installed, depending on technical and economic factors, in order to transfer thermal energy to end users without combustion inside the boiler itself.

Boilers which recover latent heat from the water vapor in stack flue gases are called condensing boilers (see also Sect. 6.8).

Table 6.3 Main properties of saturated water (liquid–vapor)

Temperature			Absolute pressure		Specific volume (m³/1,000 kg)		Enthalpy (kJ/kg)		
K	°C	°F	MPa	psi	Saturated liquid	Saturated vapor	Saturated liquid	Evaporation	Saturated vapor
273.16	0.01	32.018	0.000611	0.0886	1.0002	206,136	0.01	2,501.3	2,501.4
283.15	10	50	0.001228	0.1781	1.0004	106,379	42.01	2,477.7	2,519.8
293.15	20	68	0.002339	0.3392	1.0018	57,791	83.96	2,454.1	2,538.1
303.15	30	86	0.004246	0.6158	1.0043	32,894	125.79	2,430.5	2,556.3
313.15	40	104	0.007384	1.071	1.0078	19,523	167.57	2,406.7	2,574.3
323.15	50	122	0.01235	1.791	1.0121	12,032	209.33	2,382.7	2,592.1
333.15	60	140	0.01994	2.892	1.0172	7,671	251.13	2,358.5	2,609.6
343.15	70	158	0.03119	4.524	1.0228	5,042	292.98	2,333.8	2,626.8
353.15	80	176	0.04739	6.873	1.0291	3,407	334.91	2,308.8	2,643.7
363.15	90	194	0.07014	10.17	1.0360	2,361	376.92	2,283.2	2,660.1
373.15	100	212	0.1014	14.71	1.0435	1,673	419.04	2,257.0	2,676.1
393.15	120	248	0.1985	28.79	1.0603	891.9	503.71	2,202.6	2,706.3
413.15	140	284	0.3613	52.40	1.0797	508.9	589.13	2,144.7	2,733.9
433.15	160	320	0.6178	89.60	1.1020	307.1	675.55	2,082.6	2,758.1
453.15	180	356	1.002	145.3	1.1274	194.1	763.22	2,015.0	2,778.2
473.15	200	392	1.554	225.4	1.1565	127.4	852.45	1,940.7	2,793.2
493.15	220	428	2.318	336.2	1.1900	86.19	943.62	1,858.5	2,802.1
513.15	240	464	3.344	485.0	1.2291	59.76	1,037.3	1,766.5	2,803.8
533.15	260	500	4.688	679.9	1.2755	42.21	1,134.4	1,662.5	2,796.6
553.15	280	536	6.412	929.9	1.3321	30.17	1,236.0	1,543.6	2,779.6
573.15	300	572	8.581	1,244.5	1.4036	21.67	1,344.0	1,404.9	2,749.0
593.15	320	608	11.27	1,634.5	1.4988	15.49	1,461.5	1,238.6	2,700.1
613.15	340	644	14.59	2,116.0	1.6379	10.80	1,594.2	1,027.9	2,622.0
633.15	360	680	18.65	2,704.9	1.8925	6.945	1,760.5	720.5	2,481.0
647.29	374.14	705.452	22.09	3,203.8	3.155	3.155	2,099.3	0	2,099.3

Table 6.4 Main properties of superheated water vapor

Temperature K	°C	°F	Specific volume m³/1,000 × kg	Enthalpy kJ/kg
			$p = 0.1$ MPa $= 14.5$ psi	
			$T_{sat} = 99.63$ °C $= 211.33$ °F	
372.78	99.63	211.334	1,694	2,675.5
373.15	100	212	1,696	2,676.2
433.15	160	320	1,984	2,796.2
593.15	320	608	2,732	3,114.6
673.15	400	752	3,103	3,278.2
773.15	500	932	3,565	3,488.1
			$p = 1$ MPa $= 145$ psi	
			$T_{sat} = 179.91$ °C $= 355.84$ °F	
453.06	179.91	355.838	194.4	2,778.1
473.15	200	392	206.0	2,827.9
593.15	320	608	267.8	3,093.0
673.15	400	752	306.6	3,263.9
773.15	500	932	354.1	3,478.5
873.15	600	1,112	401.1	3,697.9
			$p = 4$ MPa $= 580$ psi	
			$T_{sat} = 250.4$ °C $= 482.72$ °F	
523.55	250.4	482.72	49.78	2,801.4
553.15	280	536	55.46	2,901.8
673.15	400	752	73.41	3,213.6
773.15	500	932	86.43	3,445.3
873.15	600	1,112	98.85	3,674.4
973.15	700	1,292	111.0	3,905.9
			$p = 16$ MPa $= 2,320.5$ psi	
			$T_{sat} = 347.44$ °C $= 657.39$ °F	
640.59	347.44	657.392	9.31	2,580.6
633.15	360	680	11.05	2,715.8
673.15	400	752	14.26	2,947.6
753.15	480	896	18.42	3,234.4
873.15	600	1,112	23.23	3,573.5
973.15	700	1,292	28.08	3,833.9

Temperature K	°C	°F	Specific volume m³/1,000 × kg	Enthalpy kJ/kg
			$p = 0.5$ MPa $= 72.5$ psi	
			$T_{sat} = 151.863$ °C $= 305.35$ °F	
425.013	151.863	305.3534	374.9	2,748.7
473.15	200	392	424.9	2,855.4
673.15	400	752	617.3	3,271.9
773.15	500	932	710.9	3,483.9
873.15	600	1,112	804.1	3,701.7
973.15	700	1,292	896.9	3,925.9
			$p = 2$ MPa $= 290$ psi	
			$T_{sat} = 212.42$ °C $= 414.36$ °F	
485.57	212.42	414.356	99.6	2,799.5
513.15	240	464	108.5	2,876.5
593.15	320	608	130.8	3,069.5
673.15	400	752	151.2	3,247.6
773.15	500	932	175.7	3,467.6
873.15	600	1,112	199.6	3,690.1
973.15	700	1,292	223.2	3,917.4
			$p = 8$ MPa $= 1,160$ psi	
			$T_{sat} = 295.06$ °C $= 563.1$ °F	
568.21	295.06	563.108	23.52	2,758.0
533.15	320	608	26.82	28,772
673.15	400	752	34.32	3,138.3
753.15	480	896	40.34	3,348.4
873.15	600	1,112	48.45	3,642.0
973.15	700	1,292	54.81	3,882.4
			$p = 32$ MPa $= 4,641$ psi	
			$T_{sat} = 400$ °C $= 752$ °F	
673.15	400	752	2.36	2,055.9
793.15	520	968	8.53	3,133.7
873.15	600	1,112	10.61	3,424.6
973.15	700	1,292	12.73	3,732.8
1,073.15	800	1,472	14.60	4,015.1
1,173.15	900	1,652	16.33	4,285.1

Fig. 6.3 Watertube boiler
basic circuit

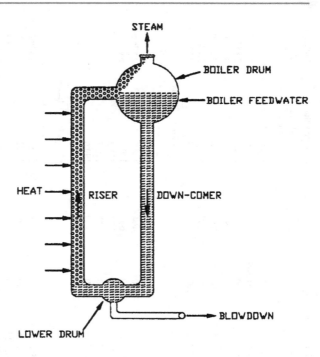

6.5.1   Steam Watertube Boilers

In watertube boilers water circulates inside the tubes and flue gas outside. The basic principle is shown in the circuit diagram in Fig. 6.3. Water circulation is generally caused by the density difference between the mixture of hot water and steam formed in the heated side of the circuit, generally called a riser, and the cooler water in the unheated side. Inside the drum, placed in the upper part of the circuit, steam is released. In normal operating conditions, water flows continuously from the drum through the cooler water downcomer and up the riser to the drum where steam is released. Industrial boilers have many parallel riser circuits carrying the steam and water mixture and few larger downcomers forming a loop.

Watertubes, like other industrial boilers, can be classified by tube shape and arrangement (horizontal, inclined, etc.), by the number of drums, capacity, or pressure.

> **Watertube units typically have capacities between 5 t of steam per hour and a hundred or more and they are able to produce high-pressure steam.**

The basic and auxiliary elements of a watertube boiler are shown in Fig. 6.4. The following points can be emphasized:

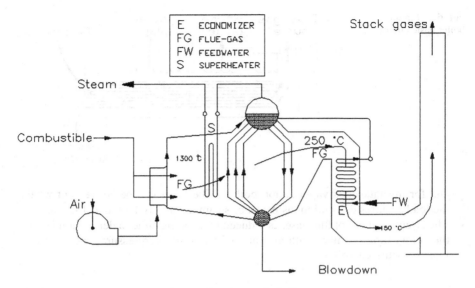

Fig. 6.4 Basic and auxiliary equipment for a watertube boiler

- In the combustion chamber fuel is introduced and burned while excess air is guaranteed by means of fans, if necessary. The heat exchange is obtained by irradiation. In this phase, where a mixture of water and vapor flows through the tubes, the heat exchange amounts to 35–40 % of the total heat exchanged inside the boiler;
- In the convective section, 40–60 % of the total heat exchanged inside the boiler is exchanged by convection. This section is the main part of the exchange surface;
- The superheater (optional) is used to produce superheated vapor. Heat is exchanged by convection and irradiation and the amount of heat is roughly 20–30 % of the total heat exchanged inside the boiler;
- The economizer (optional) is used to preheat the cold feedwater. The exchange is mainly by convection with exhaust flue gas at temperatures ranging from 673 K to 523 K (400–250 °C; 752–482 °F). It makes up roughly 5 % of the total heat exchanged inside the boiler;
- The preheater of combustion air (optional) preheats the combustion air exchanging low-temperature heat through the stack.

Notice that:

- Water circulation is generally provoked by the difference of density variation between cold feedwater and the mixture of hot water and steam in the riser;
- A high speed of the mixture inside the boiler tubes should be maintained in order to improve heat exchange. In some applications a forced circulation is used, particularly for high-pressure boilers (pressure higher than 15 MPa, 2,175 psi)

Fig. 6.5 Firetube
boiler basic circuit

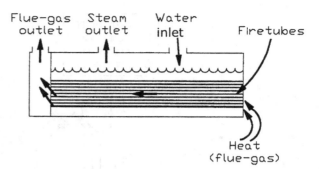

and for instantaneous-vaporization boilers. The latter can be convenient when the load requires discontinuous peaks of steam, generally less than 1–3 t/h;

• The circulation of hot flue gases is induced by creating either a depression inside the combustion chamber with aspiration fans or an overpressure with inflation fans (pressurized boilers).

6.5.2   Steam Firetube Boilers

In firetube boilers flue gas flows inside the tubes and water is outside in a shell. They are usually available with a capacity of less than 20 t of steam per hour and for pressures below 1.5 MPa (217.5 psi).

The basic and auxiliary elements of a firetube boiler are shown in Fig. 6.5.

Notice that gases make several passes, generally from 2 to 4, to increase the surface area exposed to the hot gases and to distribute steam formation more uniformly throughout the mass of water inside the shell. The main advantages of a firetube boiler include: ability to meet great and sudden load changes rapidly because of the large amount of water at the saturation temperature inside the shell, low capital cost, and low maintenance costs. The drawbacks are the thermal inertia that prolongs startup and shutdown operations and the inherent inability to superheat.

6.5.3   Heat Recovery Boilers

Heat recovery boilers belong to the class of the so-called unfired industrial boilers and are suitable for plants where process streams produce high-temperature gases which may be used to produce steam or hot water. A reliable gas supply at 573–673 K (300–400 °C; 572–752 °F) or higher is required. Of course, the fluid temperature can be raised, if necessary, by burning fuel in the heat recovery system or outside. In both cases, combustion products from the auxiliary burners are generally mixed with the hot gases before entering the convective section of the boiler; if the gas contains enough oxygen, as gas-turbine exhaust does, the auxiliary burners operate directly in the gas stream.

Watertube and firetube boilers are used as heat-recovery boilers.

Firetube boilers are suitable for low-pressure exhaust from diesel engines, kilns, furnaces, etc. and for low-capacity boilers with a capacity of less than 20 t of steam per hour.

Watertube boilers can be designed for almost any saturated or superheated steam pressure and temperature and for any capacity required. Natural or forced circulation can be used as in conventional units. Watertube boilers are generally more efficient than firetube boilers in similar conditions.

With heat recovery boilers, attention must be paid to the difference between waste gas temperature at the boiler outlet and saturated steam temperature: a small differential maximizes heat recovery, but requires a larger heat transfer area and consequently higher capital costs.

6.5.4 Electric/Electrode Boilers

Boilers fed by electricity to produce steam or hot water include two basic types: (1) resistance units where electric current flows through resistors submerged in water inside a protective tube; (2) electrode units where two electrodes are submerged in water and electric current flows through the water which has its own resistance. In this case water conductivity control is one of the main keys to good and safe operation.

Resistance units are generally designed for small systems, both hot water at low temperature and steam at low pressure.

Electrode boilers can be divided into two groups, depending on the voltage: low-voltage types operating at 600 V or lower with a maximum output of a few megawatts and medium-voltage types operating at 10–20 kV with a maximum output of 50 MW.

Both groups are available with submerged electrodes where output power is controlled by moving insulating shields or by adjusting the immersion level of the electrodes. For the medium-voltage group, spray-electrode units are available where output power is regulated by varying the amount of water sprayed upon the electrodes.

Many factors such as absence of emissions, minimal space requirements, simple maintenance, and relatively low cost may contribute to the attractiveness of these boilers, but the cost of electricity remains the main factor governing the choice.

Availability of electric power must also be taken into account because of the relatively high power needed, in comparison with that required by the whole site (roughly 1 t of steam per hour corresponds to 0.8 MW of electric power).

6.5.5   Oil Boilers

Oil and other organic fluids can be used as the heating medium to transfer thermal energy to end users for particular process demands (presses, dryers, ovens, etc.). Steam can also be generated outside the boiler by means of a hot oil/water vapor exchanger.

Forced circulation of oil is implemented by electrically driven pumps inside the tubes while flue gas is outside. The pumps' electric power and related energy consumption are relevant, although they are not included in the calculation of boiler efficiency (see Sect. 6.7).

Care must be taken to limit temperature peaks throughout the circuits, particularly if there is a power outage and pumps are switched off. Maximum temperatures of heating medium fluids are roughly 573–623 K (300–350 °C; 572–662 °F). Steam, usually at low pressure, is produced by heat exchangers outside the boiler itself.

Local regulations may accept for this system a lower control level, in both personnel and instrumentation, because there is no direct contact between flame and pressurized equipment.

6.6   Technical Parameters of Industrial Steam Boilers

The main parameters used to qualify any industrial boiler are as follows:
- Fuel type and rated consumption per hour;
- Temperature of cold feedwater and condensate;
- Temperature of combustion air;
- Flow of steam expressed in kg/h or t/h;
- Output steam pressure;
- Temperature of saturated or superheated steam;
- Temperature of flue gases from the stack;
- Design pressure which all parts of the boiler must resist. This value is generally 10 % higher than the steam pressure;
- Useful output (generally in the form of steam or hot water) expressed in kW (or practically t/h of steam).
  It follows that:

$$\text{useful output (kW)} = m \times (h - h_0)$$

where m = steam flow-rate in kg/s, h = enthalpy of the output steam in kJ/kg, h_0 = enthalpy of the input water or condensate in kJ/kg.

Typical values of $(h - h_0)$ for saturated steam output are 2,300–2,600 kJ/kg (989–1,178 Btu/lb).

Notice that the same evaluation must be carried out for a hot water or intermediate oil boiler;

- Specific capacity defined as the ratio between useful output and heat transfer area. It is expressed in kW/m^2 or kg steam/$m^2 \times$ h;

> **Boiler efficiency is expressed as the ratio between the useful output (i.e., heat in the steam) and the heat input in the fuel, based on either the Higher or the Lower Heating Value, depending on local standards (HHV in the USA; LHV in many European countries). Efficiency calculation procedures are expounded in Sect. 6.7;**
>
> **Amount of output steam produced by the unit of input fuel in rated operating conditions. Typical values for saturated steam are 12–13 kg of steam/kg of oil, 10 kg of steam/Sm^3 of natural gas.**

6.7 Boiler Losses and Efficiency

Boiler efficiency can be defined as the ratio of useful output to heat input in the fuel. Depending on local usage, the latter can be either the Higher or the Lower Heating Value.

A proper calculation of efficiency requires a definition of the boiler boundary which separates the elements to be considered part of the boiler from those that are excluded. Equipment is generally considered outside the boundary when it requires an outside source of energy (heat or electricity) or when the heat exchanged is not returned to the boiler generating system. Of course, boiler efficiency is lower than combustion efficiency (see Sect. 6.3), particularly because of additional heat losses from the exterior boiler surfaces through the insulation, generally called radiation losses.

A wider definition of boiler plant efficiency, instead of the single boiler, can take into account all kinds of energy consumption and recovery, which are not considered for the efficiency of the boiler itself.

Two basic procedures to calculate efficiency are discussed here:

1. Input–output or direct method;
2. Heat-loss or indirect method.

Of course, the two methods would give the same results if all the data required can be measured without significant error.

> **Note that efficiency is generally defined according to national rules to which reference must be made, so particular attention must be paid to fuel heat input, taken as HHV (mainly in the USA) or LHV (mainly in Europe).**

6.7.1 Input–Output Method

$$\text{Efficiency } (\%) = \left(\frac{\text{output}}{\text{input}}\right) \times 100$$

$$= \left(\frac{\text{useful output or heat absorbed by the fluid}}{\text{heat input in the fuel}}\right) \times 100$$

This method, also called the direct method, requires the direct measurement of both input energy, that is, fuel flow rate, and the useful output energy. In order to quantify the output energy, it is necessary to measure temperature, pressure, and flow rate of generated steam and boiler feedwater.

Difficulties in making these measurements, the lack of instrumentation at most industrial boiler plants, and the possibility of significant errors make large-scale use of this method impracticable.

It should be pointed out that the choice of heat input equal to either the Higher or the Lower Heating Value determines a great difference in the related efficiency values. If the ratio between the HHV and the LHV, which roughly equals 1.12 for natural gas and 1.065 for oil, is called α and the boiler output power has the same value, the ratio between the two efficiency values is:

0.893 *(for natural gas)*

$$\frac{\textbf{Efficiency (\% HHV reference)}}{\textbf{Efficiency (\% LHV reference)}} = \frac{1}{\alpha} = \frac{LHV}{HHV} =$$

0.939 *(for oil)*

6.7.2 Heat-Loss Method

$$\text{Efficiency } (\%) = 100 - \left(\frac{\text{losses}}{\text{heat input in the fuel}}\right) \times 100$$

These losses are mainly the combustion losses already discussed in Sect. 6.3. They are listed below with some comments on their origin:
- Waste heat going up the stack. The greater part is the heat carried by dry flue gases (sensible heat). Another part of the stack loss is the moisture loss (latent heat), that is, the heat used to evaporate and superheat water vapor resulting from the combustion of hydrogen in the fuel (this must be considered only if Higher

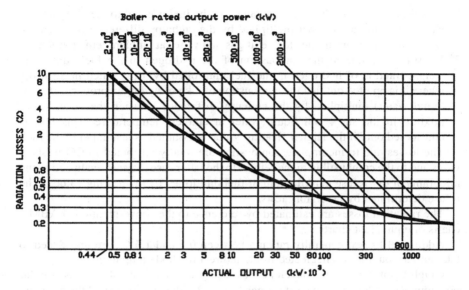

Fig. 6.6 Example of radiation losses chart (each curve corresponds to a boiler rated power value)

Heating Value is assumed as heat input in the fuel), the humidity of the combustion air, and the water contained in the fuel. Industrial boilers not equipped with waste heat recover equipment such as air preheaters or economizers have very large flue gas losses because of the high stack-gas temperature (473–573 K; 200–300 °C; 392–572 °F);

- Losses due to incomplete combustion. These consist mainly of combustible material in the flue gas such as carbon monoxide, hydrogen, and hydrocarbons. Additionally, refuse may contain unburned solid fuels and other solid combustibles;
- Radiation losses through the exterior surfaces of the boiler. Approximate evaluation of these losses by charts developed by the ABMA (American Boiler Manufacturers Association) or similar ones is shown in Fig. 6.6;
- Additional losses.

The indirect method is based on the determination of the above-mentioned losses, by first evaluating singly the heat losses per unit of fuel and then converting these values to a percentage loss by means of the heating (Higher or Lower) Heating Value.

The waste heat going up the stack is evaluated as the sum of three components:

1. Heat loss (sensible heat) due to dry gas is equal to the kilograms of dry gas per kilogram of fuel multiplied by the specific heat of the combustion gases (roughly 1 kJ/kg × K or 0.24 Btu/lb × °F) multiplied by the temperature difference between the stack exit gas (t_s) and the inlet air for combustion (t_a) The stack exit temperature t_s is closely related to the recovery equipment temperature such as economizers and air preheaters;

2. Moisture loss due to the water contained in the fuel is equal to the kilograms of water per kilogram of fuel multiplied by the enthalpy difference between the water vapor mixture in the stack exit gas and water at ambient temperature;
3. Moisture loss due to the combustion of the hydrogen in the fuel equals the weight fraction of hydrogen in the fuel multiplied by 9 (as stated in Sect. 6.1 where it is pointed out that the combustion of 1 kg of hydrogen produces 9 kg of liquid water) multiplied by the enthalpy difference between the water vapor mixture in the stack and liquid water at ambient temperature. This loss is considered only if the Higher Heating Value is assumed as heat input in the fuel.

The losses due to incomplete combustion are totally attributed to CO in stack gases and to the combustible in the refuse. In the latter case they are evaluated as the kilograms of dry refuse per kilogram of fuel multiplied by the heating value of the refuse determined by laboratory tests.

The radiation losses are estimated by reference to the above-mentioned ABMA charts or equivalent charts.

Additional losses, generally ranging between 0.5 and 1.5 %, are introduced to take into account losses neglected in the indirect method computation.

Graphic solutions were developed by using the above-mentioned or similar procedures in order to estimate stack gas losses (dry flue gas losses; all the moisture losses together) and losses due to incomplete combustion.

The heat-loss method requires the measurement of stack flue gas parameters (temperature; O_2 or CO_2 and CO concentrations as % by volume or as ppm) and of combustion air temperature.

The measurement of O_2 is generally preferred to that of CO_2 because simpler instrumentation achieves the same accuracy. Portable analyzers are available for both O_2 and CO_2 measurements.

Carbon monoxide is generally measured by means of handheld chemical absorbing analyzers (Orsat or similar analyzers) and length of stain detectors.

Stack opacity or smoke density is assumed as an index of the combustion conditions. It can be measured by means of hand pump filter paper or a similar tester where the color assumed by the paper is compared to a standard scale. Generally, index 0 means no opacity. Optimum values range between 0 and 3 (Bacharach index).

The relationships between excess air and stack gas concentration of O_2 and CO_2 for different fuels are shown in Fig. 6.7.

Figure 6.8 shows a typical relationship between total stack gas losses and stack temperature for different values of stack excess O_2 and for a specific fuel (natural gas in this case). Notice that stack temperature increases with the excess of O_2 because complete combustion is achieved. Similar curves are available for different fuels. Figure 6.9 shows the relationship between losses due to unburned combustibles and stack excess O_2 for different values of CO emissions. Sets of values referring to both Lower and Higher Heating Values are reported.

Other methods, based on the evaluation of total stack losses and unburned losses through curves and standard coefficients are widely used:

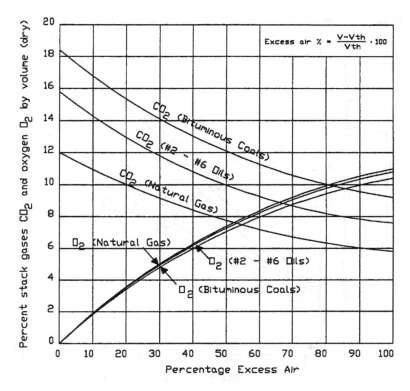

Fig. 6.7 Typical relationship between stack gas concentrations of CO_2 (%) and O_2 (%) and excess air (%)

- Total stack gas losses% $= K_s \times \frac{t_s - t_a}{CO_2\%}$ where K_s is called Hassenstein coefficient (see Table 6.5 where only the Lower Heating Value is assumed as reference), t_s is the stack gas temperature and t_a is the ambient temperature (typical values of $CO_2\%$ less than 10 %). This formula is acceptable if incombustibles are kept low;

- Losses for unburned combustible% $= K_c \times \frac{CO\%}{CO_2\% + CO\%}$ where K_c equals 50.5 for fuel oil, 37.9 for natural gas, 59 for coal if the Lower Heating Value is assumed as reference (typical values of $CO\%$ less than 0.1 %).

  Examples of evaluation are given in Sect. 6.12.

  In Fig. 6.10 the main combustion parameters are correlated by a set of curves called the Ostwald triangle. In particular, if $CO_2\%$ and $O_2\%$ are known, it is possible to check whether the combustion is complete, and if it is not, to determine the $CO\%$. Otherwise, if only $CO_2\%$ or $O_2\%$ is known, by introducing $CO\%$ (measured or estimated), it is possible to determine the other. Other sets of correlation curves also exist for use according to local practice.

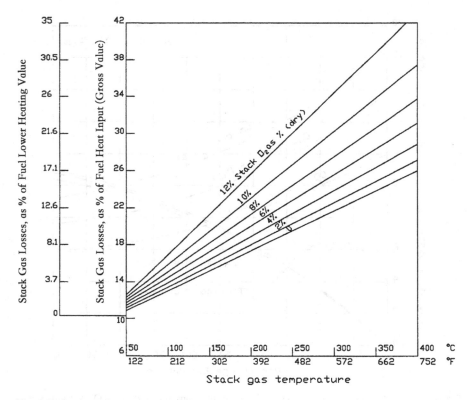

Fig. 6.8 Stack gas losses (natural gas)

Depending on heat input as fuel, either Higher or Lower Heating Value, percentage losses are as follows:

    total losses(as % of Higher Heating Value)

        = total losses(as % of Lower Heating Value)$/\alpha + 100 \times (\alpha - 1)/\alpha$

For natural gas, ratio $\alpha = 1.12$; thus

    total losses(as % of Higher Heating Value)

        = total losses(as % of Lower Heating Value)$/1.12 + 10.7$

For fuel oil ratio $\alpha = 1.065$; thus

    total losses(as % of Higher Heating Value)

        = total losses(as % of Lower Heating Value)$/1.065 + 6.1$

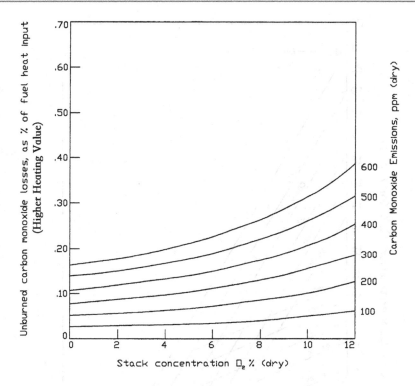

Fig. 6.9 Unburned CO losses versus O_2 concentration by volume for different values of CO emissions for natural gas (CO losses are here referred only to the Higher Heating Value; CO concentration equal to 100 ppm or 0.01 % in volume corresponds to 125 mg/Sm^3)

Table 6.5 Hassenstein coefficient values for different combustibles (K_s coefficient referred to Lower Heating Values of input fuels and to °C)

Hassenstein coefficient values[a]				
CO$_2$ % by volume	Gasoil	Oil	Natural gas	Coal
4	0.523	0.543	0.418	0.683
5	0.530	0.550	0.427	0.684
6	0.536	0.556	0.437	0.685
7	0.543	0.563	0.447	0.686
8	0.550	0.570	0.457	0.687
9	0.557	0.576	0.466	0.688
10	0.564	0.583	0.476	0.689
11	0.571	0.590	0.486	0.690
12	0.578	0.596		0.691
13	0.585	0.603		0.692
14	0.592	0.610		0.693
15				0.694

[a]If °F is used, multiply K_s by 5/9

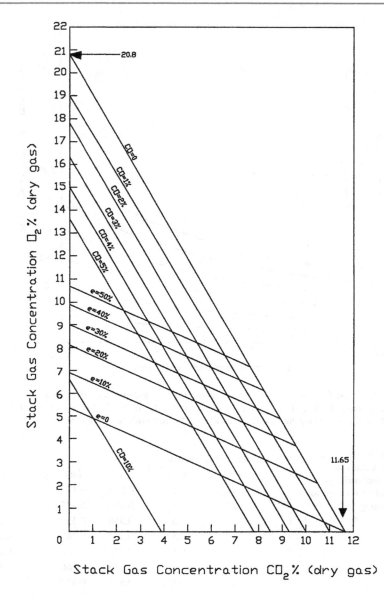

Fig. 6.10 Ostwald triangle for natural gas $e\% = \frac{V-V_{th}}{V_{th}} \times 100$; concentration by volume (dry)

6.8   How to Improve Boiler Efficiency

Boiler efficiency is influenced by many factors. As shown in Fig. 6.11, efficiency varies according to fuel, load, and the existence of stack gas recovery systems such as economizers, air preheaters, and condensing systems.

Average efficiency values range between 75 % and 90 %, and they can be obtained by a proper control of the operating parameters (the higher values within this range if recovery systems are installed and if the Net Heating Value is used as input in the fuel).

Of course, efficiency is improved when losses are kept as low as possible. This happens mainly when excess air and flue gas stack temperature are reduced.

In what follows, suggestions are made about how to minimize industrial boiler losses.

Excess air, ranging between 10 % and 60 %, but necessary to insure complete combustion and safe operation of the boiler, and high temperature of stack flue gas (473–573 K; 200–300 °C; 392–572 °F) are responsible for the most part of the stack losses.

Losses due to the latent heat of stack water vapor mixture depend on fuel and operating conditions. These losses cannot be reduced because no means are generally available to permit the water vapor mixture to condense before passing from the boiler into the stack (see condensing boilers).

Excess air can be reduced by improving the combustion control system which regulates the supply of air and fuel to meet load demand variations. Notice that a CO_2 content in stack flue gas higher or lower than standard values signals anomalous boiler operation due to lower or higher excess air values.

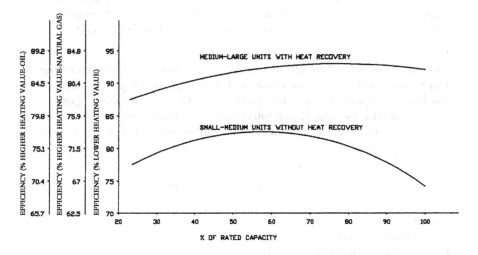

Fig. 6.11 Ranges of boiler operating efficiencies

Minimum levels of flue gas temperature are limited by potential corrosion and sulfuric acid condensation in the cold part of the stack. The minimum temperature depends on the sulfur content of the fuel and on the amount of moisture in the flue gas. As a general rule, natural gas allows for lower stack flue gas temperatures than oil fuel (around 383 K, 110 °C or 230 °F if recovery equipment is installed) because of the absence of sulfur content.

The minimum flue gas temperature is higher than that given above if recovery equipment is not installed. In this case the minimum value depends on the boiler operating pressure, which determines the saturation steam temperature. A difference of about 70–80 K or °C (126–144 °F) between the two temperatures is generally suggested. Lower values would require a larger convective surface area and consequently additional cost.

Table 6.6 shows minimum stack temperatures and minimum losses for different fuels. If stack gas temperature and excess air are higher than these values, higher losses will occur.

In condensing boilers the water vapor from the exhaust gases is turned into liquid condensate and the related latent heat is recovered. The boiler efficiency increases by 3–4 %, depending on the temperature of the cold fluid to be heated.

Losses due to incomplete combustion are generally low if the right amount of excess air is maintained. The carbon monoxide content in stack flue gas must be kept approximately equal to zero.

Boiler firing rate or output considered as a percentage of rated capacity affects efficiency, particularly at low load. Figure 6.12 shows boiler losses versus boiler firing rate in two operating conditions: (1) $O_2\%$ is kept constant over the load range (2) $O_2\%$ increases linearly as load is reduced (a common condition if combustion control is not installed). In the first situation, efficiency remains constant over a large range of load changes; in the second, efficiency diminishes as the load decreases.

In conclusion, to improve the efficiency of a boiler, excess air and stack temperature must be kept as low as possible at each load. Reduction of stack temperature is generally due to a reduction of excess air, that is, a reduction of excess O_2.

Improvement of efficiency depends on all the previous operating conditions; an improvement of about 1–2 % is generally possible even if the boiler is operating near the maximum efficiency; of course, greater improvement is possible if previous efficiency values were very low. Any efficiency improvement at constant output power will lead to energy saving as follows:

$$\text{power saving} = \left(\frac{1}{\eta_c} - \frac{1}{\eta_i} \right) \times P_{out}$$

where
η_c **= current efficiency**
η_i **= improved efficiency**
P_{out} **= constant output power**

Table 6.6 Minimum stack temperatures and losses for different combustibles without recovery equipment

Fuel	Burning equipment	C% by mass	H2% by mass	H2O% by mass	S% by mass	Others % by mass	Minimum values of stack temperature			Minimum values of excess O_2 in stack gases % by volume	LOSSES					
											Dry gas		Moisture		Total stack gas	
							K	°C	°F		(a) %	(b) %	(a) %	(b) %	(a) %	(b) %
Natural gas	All burners	75.7	23.3	0	0	1	377	104	220	1	2.9	3.2	10.1	0.	13.0	3.2
Oil N.2 (c)	All burners	87	11.9	0	0.5	0.7	439	166	330	2	5.1	5.4	6.4	0.	11.5	5.4
Oil N.6 (d)	All burners	86.6	10.8	0	1.50	0.7	472	199	390	3	6.6	7.0	6.2	0.	12.6	7.0
Bituminous coal	Pulverized	71.3	5.0	4.5	2.80	16.4	416	143	290	4	4.8	5.0	4.5	0.	9.3	5.0
Bituminous coal	Stoker	71.3	5.0	4.5	2.80	16.4	416	143	290	6	5.5	5.7	4.5	0.	10.0	5.7

(a) with reference to the Higher Heating Value
(b) with reference to the Lower Heating Value
(c) Oil N.2 according to American Society for Testing and Materials (ASTM). It is a distillate oil
(d) Oil N.6 according to American Society for Testing and Materials (ASTM). It is a residual oil

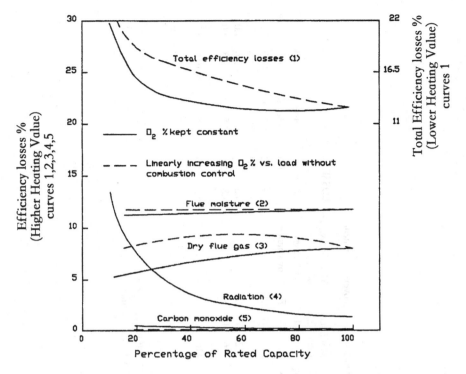

Fig. 6.12 Variation in boiler efficiency losses with changes in boiler firing rate with natural gas in two operating conditions ($O_2\%$ is kept constant to roughly 3 % in volume—see Table 6.8—over the load range by means of combustion control; $O_2\%$ increases linearly as load is reduced if combustion control is not installed)

6.9   Auxiliary Equipment

Auxiliary equipment can be divided into heat recovery systems and water treatment systems.

6.9.1   Heat-Recovery Systems

Flue gas going up the stack contains a considerable amount of thermal energy, part of which can be recovered by installing air preheaters and economizers.

Air preheaters are used to transfer heat from the stack flue gas to the incoming combustion air. In this way boiler efficiency is improved because of (1) reduction of stack gas temperature (2) reduction of excess air due to improved burner combustion conditions if preheated air is supplied.

With air preheaters the boiler's total efficiency will increase significantly: roughly 2 % for every 50 K or °C (90 °F) decrease in temperature of flue gas or 2 % for every 50 K or °C (90 °F) increase in that of combustion air.

A high combustion air temperature seems to increase the emission of nitrogen oxides, particularly in the case of natural gas. The reduction of excess air, however, may moderate the NO_x increase.

Maximum combustion air temperature does not generally exceed 200 °C (392 °F), depending on fuel, type of burner, and boiler condition. Higher temperatures are required if pulverized coal is burned. The lowest flue-gas temperature at the outlet of the preheater is limited by potential corrosion, as mentioned in Sect. 6.8.

Air preheaters can be classified as either recuperative or regenerative.

In recuperative units, heat is transferred directly from the hot medium such as hot flue gas on one side of the stationary surface to air on the other side.

Regenerative units transfer heat indirectly from the hot flue gas to combustion air through an intermediate heat storage medium. They are used mainly in retrofitting existing boiler plants.

An alternative to the above-mentioned preheaters is the heat-pipe heater (see Sect. 15.5). They are not currently in use in boiler units, but they may become competitive with the traditional ones.

Economizers provide additional heating to the feedwater by recovering heat from stack flue gas. In this way the boiler's firing rate needed to produce the same amount of steam is reduced and the overall efficiency is increased. Economizers can also be installed to heat hot water for users apart from the boiler water cycle.

The advantage of economizers is closely related to the existing feedwater temperature and condensate return system. Roughly, an increase in feedwater temperature of 5 K or °C (9–10 °F) will result in an efficiency increase of 1 %.

In case of significant blowdown (over 10 % of output steam) or of low condensate recovery, heat recovery can be installed to preheat the incoming feedwater.

6.9.2   Water Treatment Systems

Water treatment is one of the main factors in boiler reliability. It is necessary because impurities in the feedwater can lead to scale deposits on the boiler heating surfaces and to corrosion. Suspended solid matter, colloids, dissolved minerals, atmospheric gases, and impurities coming from the steam cycle are always present in varying degrees in boiler feedwater.

The main equipment for water treatment can be summarized as follows: demineralization, pretreatment systems to remove suspended solids such as clarifiers, softeners, and media and precoat filters.

Demineralization removes dissolved salts from pretreated water, where they are present in the form of positive and negative ions (cations and anions). The goal is reached by means of ion-exchange resins exchanging harmless ions with impurities such as sodium cations and chloride anions. Exhausted resins can be regenerated for reuse; after a certain number of regenerations the resin bed must be discarded. The two-column system, which is one of the commonest installations, includes an acid cation exchanger followed by a base anion exchanger with a degasifier interposed to strip out the CO_2 produced in the first exchanger. Boilers operating at high pressure and temperature require feedwater of very high purity (roughly 5–10 parts per billion of sodium and silica). In this case other systems such as mixed cation–anion-resin beds and semipermeable membranes are generally installed.

Filters and clarifiers are used to remove solid particles and some dissolved substances. Clarifiers work with chemical coagulants which help to gather finely divided matter into larger masses. In addition, calcium and magnesium can precipitate depending on the chemicals added. Calcium and magnesium hardness can also be removed by separate softeners.

Softeners are used to remove hardness and also to remove silica and reduce carbonate alkalinity. They generally work at ambient temperature and are effective for low-pressure boilers that do not require complete feedwater demineralization. Hot-process softeners operating at above 373 K (100 °C; 212 °F) can be used for higher-pressure boilers.

Notice that blowdown is commonly used to remove boiler water impurities. The amount depends on many factors, particularly on boiler water treatment performances, and it can be as much as 5–10 % of the total boiler steam. It is always possible to use heat recovery from blowdown water to preheat the incoming feedwater water. Low-pressure boilers generally use heat exchangers; high-pressure and high-capacity boilers use flash systems where part of the blowdown wastewater is flashed into high-purity steam, then condensed and recycled into the feedwater system. The remaining waste water exchanges heat with the makeup water and then is discarded.

Table 6.7 synthesizes the main features of the commonest types of water treatment equipment.

6.9.3   Condensate Return Systems

The recovery of condensate from process steam furnishes a great amount of usable energy and clean feedwater. The degree of recovery and its economic validity depend on many factors such as contamination from process, layout of steam end users and pipelines. The quantity of energy contained in the saturated liquid can be estimated as roughly 20 % of the original energy contained in the steam at the same pressure.

Table 6.7 Effect of treatment techniques on makeup water contaminants under ideal operating conditions

Treatment techniques	Suspended Solid	Alkalinity	Hardness	Dissolved silica	Dissolved solids
Filtration	Excellent	No change	No change	No change	No change
Cold-process softening	Good	Fair	Fair	Good	Fair
Hot-process softening	Good	Good	Good	Good	Fair
Sodium-cycle cation exchange	Excellent	No change	Excellent	No change	No change
Demineralization	Excellent	Excellent	Excellent	Excellent	Excellent
Reverse osmosis	Excellent	Excellent	Excellent	Excellent	Excellent

The main systems used to recycle the condensate into the boiler are: the atmospheric system, pressurized systems, and deaerators.

The atmospheric system consists of an open tank where condensate is received downstream of steam traps or drains. At atmospheric pressure the condensate temperature is reduced to 373 K (100 °C; 212 °F): a portion of the condensate, about 10–15 %, will flash to steam and will be lost in the atmosphere; the remaining hot water will feed the boiler together with as much treated makeup water added as the boiler requires.

The semi-pressurized and fully pressurized systems work with a receiver at intermediate pressure below the lowest process steam pressure or at process steam pressure. The flashed steam is piped to a low-pressure steam main where it can be utilized for process purposes; the lost flashed steam is conveyed to the outlet through relief valves or automatically operated vents which are installed to prevent abnormal pressure surges. This loss, too, can be reduced if some additional provisions are made for steam recovery.

The condensate is generally cooled by the makeup water to at least 283 K (10 °C; 18 °F) below the saturation temperature to prevent vapor generation in the boiler feedwater pump system.

Notice that although the use of flash steam in low-pressure mains seems to offer a significant heat recovery, its practical application involves a number of problems, essentially economic, that must be carefully considered. In addition to inherent factors which must be taken into account when planning a condensate recovery system, the quantity of flash steam must match potential needs at any time since steam cannot be stored economically for later use.

Spray deaerators, also called barometric condensers, consist of a tank where steam rises through a cold water spray which condenses it (see Fig. 6.13). This system, which achieves complete flash steam recovery to heat boiler feedwater, has the additional advantage of producing a deaerating effect: if the temperature of the tank is kept above about 363 K (90 °C; 194 °F) by a proper metering of the spray, dissolved gases in the condensate and feedwater, such as oxygen and CO_2, will come out of the solution and will be released through the atmospheric vent.

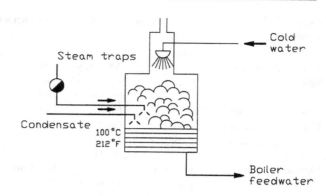

Fig. 6.13 Flash steam recovery in spray tank

6.10   Computerized Control of Boiler Combustion

Controlling the boiler plant means ensuring a correct balance of energy and mass entering and leaving the boiler.

Fuel is the major energy input, and combustion air is the major mass input. The only useful energy output is steam (or hot water or other intermediate fluids), although heat from flue gas is often recovered.

A combustion control system must regulate the quantity of fuel and combustion air to meet the load requirements automatically by ensuring minimum fuel consumption, minimum pollution, and safe operation at the same time. Flue-gas analyzers and measurements of input combustion air and fuel flow, and of the energy output parameters (steam pressure, temperature, and flow) give all the information about boiler performance that a boiler control system may need.

The complexity of the control system is generally related to the capacity of the boiler plant and to the legal requirements for safe operation and air pollution.

The simplest system, once extensively applied to industrial boilers, consists of a single actuator. When the steam pressure changes in consequence of a change in load demand, this actuator moves both the fuel and combustion air input devices, mechanically linked, to a preset position indicating a preset value of the fuel–air ratio. This system cannot compensate any shift from design conditions, so that high excess-air levels are generally preset to guarantee safe boiler operation. Thus abnormal excess air may be present throughout boiler life.

Separate actuators for the fuel valve and the air combustion fan, together with an individual manual adjustment of the fuel–air ratio, can lower the margin of excess air and consequently improve boiler efficiency. An operating guide, based on the availability of oxygen analyzers or the steam flow–air flow ratios, must be accessible to the operator so that he can optimize the amount of excess air.

Metering of fuel and airflow by means of pressure measurements or directly by means of flow metering is a further improvement in boiler control. This is called cross-limited metering control system and it consists in linking fuel flow to the airflow through a control logic.

A cross-limited metering control by continuous monitoring of flue gas O_2 levels can eliminate the effects of variations in fuel heating values and combustion air conditions. The control system can also be preset with an O_2 level corresponding to minimum fuel consumption. Oxygen analyzers using zirconium oxide sensors and computers are currently used for reliable operation.

6.11 Boiler Emission and Environmental Pollution

Emission control is undoubtedly one of the main requirements in boiler plant operation. Depending on fuels and boiler equipment, exhaust gases contain by-products in different quantities, some of which are air pollutants. Small and medium-sized boiler plants, which cannot economically accommodate abatement systems, must cope with this problem by a correct choice of fuel and/or of burners and auxiliary equipment.

Although every country has its own environmental protection regulations, a general classification of major pollutants can be made as follows: sulfur dioxide, nitrogen oxides, carbon monoxide, particulate matter, and hydrocarbons. In addition, other pollutants are always present depending on the particular fuel and operating conditions.

Typical ranges of values of permitted pollutant concentrations in air emission from boiler plants are reported in Table 6.8 for the above-listed pollutants. The permitted values vary from country to country, but boiler plants must be conducted so as to ensure values lower than those in the table.

6.11.1 Sulfur Dioxide

Sulfur dioxide SO_2 is formed during combustion (see Sect. 6.2) by the combination of sulfur contained in the fuel with oxygen from combustion air.

Table 6.8 Typical ranges of values of permitted pollutant concentrations from combustion

Pollutants	Level of concentration C_R[a]	
	mg/Sm^3	ppm[b]
Sulfur dioxide	1,500–2,000	525–700
Nitrogen oxides[c]	50–200	102–410
Carbon monoxide	100	80
Particulate matter	100–150	

[a]Referred to dry gas and to a concentration of O_2 equal to 3 % in volume. To convert from the actual pollutant concentration C_x (with an oxygen concentration W_x) to the concentration referred to 3 % value of oxygen concentration, use the following relationship: $C_R = (18/(21 - W_x)) \times C_x$
[b]10,000 ppm = 1 % in volume
[c]Expressed in NO_2

Except for sulfur compound particles, most of the sulfur contained in the fuel is converted into SO_2, but a small percentage of the total oxides is converted into sulfur trioxide SO_3. Sulfur oxides combine with the moisture present in stack flue gas to form sulfuric acid, some of which can condense and corrode metallic parts of the boiler. Similarly, in the atmosphere a portion of SO_2 is converted into SO_3 and then into sulfuric acid and sulfur compounds.

The quantity of sulfur oxides formed during combustion is closely related to the sulfur content of the fuel and does not depend on boiler operating conditions. Thus, reduction of sulfur oxides can be obtained by using low-sulfur fuel, natural gas in particular, or by introducing abatement devices such as stack gas scrubbers, which remove SO_2 from stack flue gas.

The maximum SO_2 concentration in stack gas must be kept lower than 1,500–2,000 mg/Sm^3 or 525–700 ppm (see Table 6.8).

Emulsifying fuel oil with water or other compounds and atomization at the burner are other techniques widely applied to solve problems due to the use of heavy oil.

6.11.2 Nitrogen Oxides

Nitrogen oxides NO_x are formed during combustion by the combination of oxygen and nitrogen at high temperature. Both components are naturally present in combustion air and nitrogen in the fuel itself.

NO_x emissions are mainly NO (95 % in mass) and NO_2 (5 % in mass). In the atmosphere NO combines with oxygen to form NO_2.

NO_x formed during the combustion process can be classified as Thermal, Fuel, or Prompt:

- Thermal NO_x (via Zeldovich mechanism) is formed due to the oxidation of atmospheric nitrogen at temperatures higher than 1,273 K (1,000 °C, 1,832 °F). It becomes important when the temperature reaches the range 1,600–1,800 K (1,327–1,527 °C, 2,421–2,781 °F);
- Fuel NO_x is formed due to the oxidation of the nitrogen in the fuels (its value ranges between 0.05 and 1.5 % in mass depending on the kind of fuel; lower values for natural gas and light oil, higher values for heavy oil and coal). The oxidation is accelerated if excess air value is high;
- Prompt NO_x derives from the reaction of atmospheric air with hydrocarburic radicals.

The concentration of NO_x in stack gases is related mainly to the quantity of nitrogen in the air; typical values are 50–200 mg/Sm^3 or 102–410 ppm (see Table 6.8).

Table 6.9 Main techniques used to reduce NO_x emissions

Techniques		Effects on NO_x emissions	Combustion efficiency
Excess air	(+)	(+)	(−)
Flame temperature	(+)	(+)	(+)
Preheating of air	(+)	(+)	(+)
Steam injection	(−)	(−)	
Gas recirculating	(+)	(−)	
Combustion chamber size	(+)	(−)	
Combustion chamber charge	(+)	(−)	
Heat production rate	(+)	(+)	
Heat production exchange	(+)	(−)	
Duration of combustion	(+)	(+)	
Burners integration	(+)	(+)	
Nitrogen in fuel	(+)	(+)	
Type of fuel			
Coal	(+)		
Oil	(+)		
Natural gas	(−)		

(+) means increase, (−) means decrease

Although the reduction of nitrogen in the fuel is a technique that should be practiced, most technologies applied to reduce NO_x are based on lowering peak flame temperature and on reducing the amount of oxygen available in the flame. Low excess-air burners, fuel-rich or staged-firing burners with separate primary and secondary combustion zones and flue gas recirculation are the main means used to reduce the production of NO_x inside the boiler. In the case of coal, fluidized-bed boilers are suitable for this purpose. Steam injection in the combustion chamber is also used to reduce NO_x.

Post-combustion NO_x can be reduced by means of abatement devices generally installed downstream of the combustion zone but above air preheaters such as selective-catalytic-reduction units (SCR) and selective-non-catalytic-reduction units (SNR). The choice of the best solution, both technically and economically, must be made individually for each plant in the light of local regulations and boiler operating modes.

Table 6.9 lists the main techniques used and comments on how they can reduce NO_x emissions and influence combustion efficiency.

6.11.3 Carbon Monoxide

Carbon monoxide CO is always produced with incomplete combustion (see Sect. 6.2), and its concentration in stack flue gas is closely related to boiler operating conditions. CO concentration at the stack must be very low; concentrations higher than 0.5 % signal poor combustion conditions and generally low excess air. CO emitted from the stack is dispersed into the atmosphere where it adds to that from other sources of CO such as internal combustion engine vehicles.

> **The maximum CO concentration in stack gas must be kept lower than 100 mg/Sm^3 or 80 ppm (see Table 6.8).**

6.11.4 Particulate Matter

Particulate matter includes a wide variety of materials such as unburned fuel, sulfur compounds, carbon, ash, and non-combustible dust that enter the combustion chamber with combustion air.

The quality and quantity of particulate matter are influenced mainly by the type of fuel, the boiler's operating mode and the type of burner. Natural gas and some light oil fuels produce little solid matter and ash. Most coals and heavy oils produce a great quantity both of particulates in stack flue gas and of ash, some of which remains in the stack flue gas.

In the case of coal, ash can amount to 20 % or more of the total weight of the fuel. Particular attention must be paid during both the design and the maintenance of the boiler to avoid the accumulation of ash on internal boiler surfaces.

To prevent particulate concentration from exceeding emission standards and to reduce health hazards, various abatement techniques are used, such as filtration, mechanical separation, and electrostatic precipitation. These techniques can be applied to both coal and heavy oil boiler plants.

> **The maximum particulate matter concentration in stack gas must be kept lower than 100–150 mg/Sm^3 (see Table 6.8) depending on the quality of the fuel.**

6.11.5 Hydrocarbons

Hydrocarbons can be grouped as either unburned fuel components or compounds from chemical reactions occurring during combustion.

Hydrocarbons can be reduced by proper combustion, but traces of their compounds will always be present in stack flue gas.

6.11.6 Other Pollutants

Other pollutants can be found in stack flue gas depending on the type of fuel.

With heavy fuel oils, in addition to the pollutants already mentioned, asphaltenes can form a finely dispersed colloid producing tar particles which deposit in storage tanks and travel through the burner without volatilizing. Asphaltenes can be stabilized by chemical additives; the same technique can be applied to prevent formation of vanadium pentoxide which is known to promote sulfuric acid reactions.

When refuse-derived fuels (RDF) are used, the stack flue gas contains particulates, complex hydrocarbons, trace metal emissions, and chlorides such as HCl, depending on both the nature of the refuse and the kind of treatment upstream the combustion for separating and/or recycling part of the waste.

6.11.7 Carbon Dioxide and Greenhouse Gases

Carbon dioxide, which is naturally found in the atmosphere as part of the earth's carbon cycle, is the primary greenhouse gas emitted through human activities. These are mainly electricity production, transportation, and industrial processes involving fossil fuel combustion and other chemical reactions.

In fuel combustion carbon dioxide emission is strictly related to combustion efficiency: the higher the efficiency, the greater the quantity of carbon dioxide emission.

> **Since greenhouse gases are considered one of the main causes of climate change, many countries have set regulations to limit the maximum amount of CO_2 that each site can produce.**
>   **Worldwide accepted indicators are:**
>
> | | | |
> |---|---|---|
> | 2 kg CO_2/Sm^3 natural gas | 3.2 kg CO_2/kg oil | 2.4 kg CO_2/kg coal |
> | 32.04 lbCO_2/Sft^3 natural gas | 3.2 lb CO_2/lb oil | 2.4 lb CO_2/lb coal |

Notice that the most important GHG (greenhouse gas) emission due to human activity includes CO_2, CH_4 (methane), NO_2 (nitrous oxide), water vapor, O_3 (tropospheric ozone), and the so-called F-gases that are often used in insulations, foaming materials, fire extinguishers, solvents, pesticides, and aerosol propellants.

F-gases include hydrofluorocarbons (HFCs), perfluorocarbons (PFCs), and sulfur hexafluorides (SF6). These gases are mainly used as substitutes for ozone-depleting substances such as chlorofluorocarbons (CFCs), hydroclorofluorocarbons (HCFCs), and halons.

For the application of these gases as coolants in cooling plants, see Sect. 12.3.

6.12   Practical Examples

Two examples of efficiency evaluation for industrial boilers are presented here below with reference to both the Higher and the Lower Heating Value of the fuel.

Example 1 Calculate the efficiency of an industrial watertube steam boiler by means of the heat-loss method.

This concerns an industrial watertube steam boiler, fed by natural gas, with the technical characteristics stated below. The efficiency evaluation has been conducted by the indirect method and by measuring $O_2\%$ and $CO\%$. Correlation curves and relationships introduced in Sect. 6.7 are here used to determine $CO_2\%$ and losses (Ostwald triangle, K_s and K_c coefficients). The radiation losses are estimated by a chart similar to the chart in Fig. 6.6.

Nominal boiler operating conditions	
Steam flow	18 t/h
Final saturated steam pressure	7 MPa = 1,015.2 psi
Condensate return temperature	70 °C = 158 °F
Rated output m × ($h - h_0$)	12,350 kW = 42,136 Btu/h
Test operating conditions	
Steam flow (estimated)	16 t/h
Stack temperature (measured)	148 °C = 298 °F
Air temperature (measured)	28 °C = 82 °F
$O_2\%$ (measured)	4.6 %
CO% (from Ostwald triangle, see Fig. 6.10)	0.013 %
$CO_2\%$ (measured)	9.2 %

Efficiency evaluation (Lower Heating Value as reference 34,325 kJ/Sm³)
• Total stack gas losses

$$K_s \times \frac{t_s - t_a}{CO_2\%} = 0.468 \times \frac{148 - 28}{9.2} = 6.1\%$$

• Losses due to unburned combustible

$$K_c \times \frac{CO\%}{CO_2\% + CO\%} = 37.9 \times \frac{0.013}{9.2 + 0.013} = 0.05\%$$

• Radiation losses (see Fig. 6.6)
    1.2 %

$$\text{Efficiency} = 100 - (6.1 + 0.05 + 1.2) = 100 - 7.35 = 92.65\%$$

Efficiency evaluation (gross heating value as reference 38,450 kJ/Sm³)
• HHV/LHV = 38,450/34,325 = 1.12
• Total losses referred to the Higher Heating Value if the output power is kept constant

$$(6.1 + 0.05 + 1.2)/1.12 + (112 - 100)/1.12 = 17.29\%$$

$$\text{Efficiency} = 100 - 17.29 = 82.71\%$$

Efficiency is definitely high, so no improvement is practicable.

Example 2 For a given rated output power, compare the energy consumption of two steam boilers having different efficiencies.

This concerns an industrial firetube steam boiler, fed by natural gas, with the technical characteristics detailed below.

The efficiency evaluation has been conducted by the indirect method and by measuring $O_2\%$, $CO_2\%$, and $CO\%$. Correlation curves and relationships introduced in Sect. 6.7 are here used to determine losses (K_s and K_c coefficients). Radiation losses are estimated by a chart similar to the chart in Fig. 6.6.

Nominal boiler operating conditions	
Steam flow	10 t/h
Final saturated steam pressure	0.9 MPa = 130.5 psi
Condensate return temperature	80 °C = 176 °F
Rated output $m \times (h - h_0)$	6,805 kW = 23,217 Btu/h
Test operating conditions	
Steam flow (measured)	8 t/h
Stack temperature (measured)	245 °C = 473 °F
Air temperature (measured)	21 °C = 69.8 °F
$O_2\%$ (measured)	5.2 %
$CO\%$ (for Ostwald triangle, see Fig. 6.10)	0.01 %
$CO_2\%$ (measured)	8.7 %

Efficiency evaluation (Lower Heating Value as reference 34,325 kJ/Sm³)
- Total stack gas losses

$$K_s \times \frac{t_s - t_a}{CO_2\%} = 0.467 \times \frac{245 - 21}{8.7} = 12.02\%$$

- Losses for unburned combustible

$$K_c \times \frac{CO\%}{CO_2\% + CO\%} = 37.9 \times \frac{0.01}{8.7 + 0.01} = 0.043\%$$

- Radiation losses (see Fig. 6.6)
  1.8 %

$$\text{Efficiency} = 100 - (12.02 + 0.043 + 1.8) = 100 - 13.86 = 86.14\%$$

Efficiency evaluation (Higher Heating Value as reference 38,450 kJ/Sm³)
- HHV/LHV = 38,450/34,325 = 1.12
- Total losses referred to the Higher Heating Value if the output power is kept constant

$$(12.02 + 0.043 + 1.8)/1.12 + (112 - 100)/1.12 = 23.09\%$$

$$\text{Efficiency} = 100 - 23.09 = 76.9\%$$

Table 6.10 Comparison between two boilers with different efficiencies (Example 2) fed by natural gas (34,325 kJ/Sm^3 as Lower Heating Value))

	10 t/h			2,770 kJ/kg	Saturated steam	
6,805 kW	0.9 MPa = 130.5 psi			320 kJ/kg	Condensate	
Output	Working hours		Efficiency[a]	Natural gas input		
% rated capacity (%)	kW	h/year	%	kW	Sm^3/h	Sm^3/year
Existing boiler-rated steam flow						
80	5,444	2,000	86.13	6,321	663	1,325,823
40	2,722	2,000	80	3,403	356.9	713,707
20	1,361	2,000	76	1,791	187.8	375,635
Total (a)						2,415,165
New boiler-rated steam flow						
80	5,444	2,000	91	5,983	627.5	1,254,870
40	2,722	2,000	90	3,025	317.2	634,406
20	1,361	2,000	88	1,547	162.2	324,412
Total (b)						2,213,688
Energy saving (a) − (b)					Sm^3/year	201,477
					TOE/year	165.210

[a]LHV as reference 34,325 kJ/Sm^3

The efficiency can be increased up to 91–92 % (referred to LHV) or to 81–82 % (referred to HHV) by replacing the boiler (particularly if it is near the end of its useful life) or by installing proper equipment to improve combustion efficiency or heat recovery.

Table 6.10 shows energy-saving evaluations for different operating conditions over a year.

The economic evaluation is in Table 19.4.

Electric Distribution Systems from Facilities to End Users

7

7.1 Introduction

Electric energy, which is generally delivered to sites by means of utility electric networks at high or medium voltages, is distributed to end users in the site through medium- and low-voltage networks. Depending on the power demand of loads and on the process layout, transformer substations (see Chap. 5), whose main task is to reduce the supply voltage downstream of the utility-delivering node, are located at the site boundary or distributed around the site itself.

The choice among different distribution systems, such as radial or loop-feeder systems, shown in the diagram in Fig. 7.1, is based on technical and economic evaluations that generally do not consider energy-saving targets, because of the insignificant amount of the energy losses involved.

> As a general indication, only a small part of the total electric power travelling along the internal lines, not more than 2–3 %, is lost as Joule and additional losses, both related to the square of the flowing current. If compared to the heat distribution losses that can reach 10–20 % of the total thermal energy flowing along pipelines (see Chap. 8), this figure shows that these problems can easily be regarded as of secondary importance in an energy management program.

Most electric losses occur in end users which can be grouped basically as (1) electrical machinery and drives; (2) electrically heated users such as furnaces, ovens, boilers, induction heating equipment, resistors, and microwave equipment; (3) lighting; and (4) others such as electrochemical equipment and control and communication systems.

G. Petrecca, *Energy Conversion and Management: Principles and Applications*,
DOI 10.1007/978-3-319-06560-1_7, © Springer International Publishing Switzerland 2014

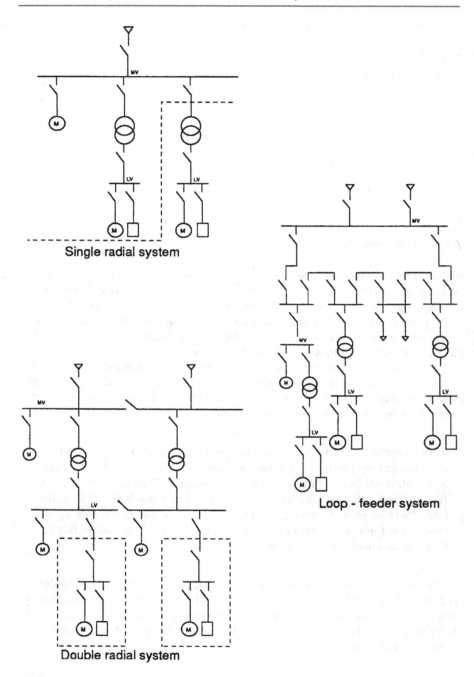

Fig. 7.1 Industrial electric distribution systems: Radial, loop feeder

Electrical drives and electrically heated users are discussed in this chapter; lighting will be the topic of Chap. 14; for electrochemical and other applications, it is advisable to consult specialized technical literature.

7.2   Electric Distribution Losses

Electric distribution losses in the site are mainly due to the Joule effect which depends on the square of the current and on the line's resistance. The basic relationship can be expressed as follows:

$$P_{\text{losses}} = n \times R \times I^2 \ (\text{W})$$

where

n = number of phase conductors.
R = resistance of the phase conductor (Ω).
I = RMS or effective current (A).

If the current flows through a series of n electric conductors, the total resistance is expressed as the sum of the single conductor's resistances:

$$R_{\text{total}} = \sum_1^n j\rho_j \times \frac{l_j}{S_j} \ (\Omega)$$

where

ρ_j = resistivity of the conductive material which is defined as resistance per unit
      length of unit area at a given temperature (reference or operating temperature).
S_j = cross section of the j-conductor.
l_j = length of the j-conductor.

In the case of a line having n electric conductors connected in parallel sharing the total current, the lower the resistance the higher the current flowing in each conductor is. The total resistance is expressed as the reciprocal of the sum of the reciprocals of the resistances:

$$R_{\text{total}} = \frac{1}{\sum_1^n j \frac{1}{\rho_j} \times \frac{1}{l_j} \times S_j}$$

In practice, if n conductors are in parallel and they have the same section, the resistance of the line is $(1/n)$ multiplied by the resistance R of each conductor:

$$R_{\text{total}} = \left(\frac{1}{n}\right) \times R$$

Table 7.1 Technical parameters for different conducting materials

Parameter		Silver	Copper	Aluminum	Lead
Density[a]	kg/m^3	10,523	8,937	2,707	11,372
	lb/ft^3	657	558	169	710
Resistivity[a]	Ω m	1.60×10^{-8}	1.76×10^{-8}	2.83×10^{-8}	2.10×10^{-7}
	Ω mm^2/km	16.0	17.6	28.3	210
	%/°C	0.38	0.39	0.40	0.40
Temperature coefficient					
	%/°F	0.21	0.22	0.22	0.22
Specific heat[a]	kJ/kg K	0.234	0.383	0.896	0.130
	Btu/lb °F	0.056	0.092	0.214	0.031
Melting point	°C	961	1,083	658	327
	°F	1,762	1,981	1,216	621
Thermal conductivity[b]	W/m °C	415	381	228	33
	Btu/h ft °F	240	220	132	19

[a]At 20 °C = 68 °F
[b]At 100 °C = 212 °F

> **Industrial distribution systems are mainly three-phase systems; here, too, the relationships shown above are used as follows:**
>
> $$P_{losses} = 3 \times R \times I^2$$
>
> **where**
>
> **R = resistance of a single-phase conductor.**
>
> **I = RMS or effective value of the current flowing in each line.**

The resistivity of conducting materials, such as copper and aluminum, varies with the temperature at a rate of about 0.4 %/°C (0.22 %/°F). The reference temperature for resistivity values is 75 °C (167 °F) for electrical machinery and 20 °C (68 °F) for cables. The temperatures at which cables can work in steady-state conditions range from 50 to 100 °C (122–212 °F); this range derives from the ability of various insulating materials to support high temperatures for a long time without unacceptable deterioration.

Additional losses occur if the frequency of the current differs from zero. In this case, which is the commonest, since losses vary according to the square of the current as do the basic Joule losses, an equivalent resistance is generally introduced to take these phenomena into account. In practice, this resistance is expressed as a percentage of the zero frequency or direct current resistance; values range from 3 to 100 % and higher. As a general rule, the higher the frequency and the larger the conductor's cross section, the higher the percentage is. By their nature additional losses decrease as the temperature rises.

Values of resistivity for different materials and other technical parameters are reported in Table 7.1. Note that the resistivity is here expressed also in Ω mm^2/km in order to facilitate the calculation for line conductors.

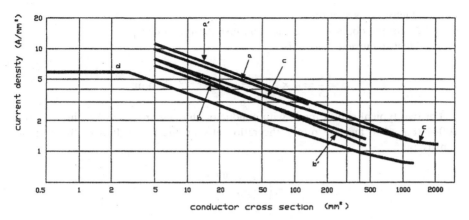

Fig. 7.2 Current density of different copper cables: (*a*) Single-core paper insulated in air; (*a'*) single-core paper insulated underground; (*b*) three-core paper insulated in air; (*b'*) three-core paper insulated underground; (*c*) non-insulated conductor in air; (*d*) single-core rubber insulated cables

The main parameters to consider in designing electrical systems are current density and voltage drop through lines and transformers, to which losses are related. Widely accepted criteria for the design of electric power distribution systems in industrial plants are based on a predetermined voltage drop along the lines, generally fixed at less than 5 % in normal operating conditions, to which corresponds less than 2–3 % of losses along the same line. These values are related to the ratio (resistance *R*/reactance *X*) of the cable and to the power factor of the load. The maximum current density in this situation depends on operating temperature and conductor cross section. As a general rule, the larger the cross section of the conductor the lower the acceptable current density is (see Fig. 7.2).

For low-voltage cables (0.6–1 kV), and for various modes of installation, the following formula may be used to calculate electric current capacity:

$$I = A \times S^m - B \times S^n \text{ (A)}$$

where *A*, *B*, *m*, and *n* are parameters defined by IEC Standard and *S* is the conductor's cross section.

A simplified formula for non-buried cables is

$$I = a \times S^b \text{ (A)}$$

where
$a =$ current capacity of a conductor with a cross section of 1 mm^2. Typical values for copper conductors are 13 A/mm^2 for PVC insulation and 17 A/mm^2 for EPR, XLPE insulation.

S = conductor cross section in mm^2.
b = 0.625 for the commonest cables and installation modes.
The electric current density is

$$\delta = \frac{I}{S} = a \times S^{b-1} = a \times S^{-0.375} \ \text{A/mm}^2$$

For two typical cables made with conductors having cross sections in the ratio 100–1 (e.g., 150 and 1.5 mm^2), the ratio between the two values of the current density is

$$\frac{\delta_{150}}{\delta_{1.5}} = \left(\frac{150}{1.5}\right)^{-0.375} = 0.17$$

This example shows that a given current density cannot be assumed as reference value to design cable lines, but that voltage drops and continuous operating temperatures are the main parameters to be considered.

Notice that lines are generally exploited less than they have been designed to, because a site rarely expands as quickly as foreseen; in consequence, line losses are much lower than the values given above.

In conclusion, because actual losses are comparatively small, energy-saving design criteria are not extremely important for the design or the retrofit of the industrial line.

7.3   Power Factor Control

The power flowing through the electric lines has two components: (1) the active power (kW), which is the power available for conversion to mechanical, thermal, chemical, light, or sound energy, and (2) the reactive power (kvar), which is used for exciting magnetic fields in transformers and electrical machinery or electric fields in capacitors.

The expressions for the active and reactive power are as follows:

$$P = \sqrt{3} \times V \times I \times \cos \varphi \ \text{(kW)}.$$
$$Q = \sqrt{3} \times V \times I \times \sin \varphi \ \text{(kvar)}.$$

The power factor, i.e., the cosine of the angle between voltage and current in a specific section of an electric circuit, is an index of the ratio between the active and reactive power flowing in the same section:

$$\tan \varphi = Q/P$$
$$\cos \varphi = \cos \ \text{atan} \ Q/P$$

This relationship can be used to calculate the power factor value in each section of the network.

In order to minimize losses in an existing line, the current flowing along it must be kept as low as possible while ensuring that the active power P required by the loads can still flow.

A constant amount of power P at constant voltage can be carried by keeping the product $I \times \cos \varphi$ constant. The higher the power factor $\cos \varphi$, the lower the current will be and in consequence line losses.

In electric circuits, this means that electric active power is transferred with the reactive power intrinsically needed by the majority of electric loads such as motors and transformers. If the reactive power is fed in nodes along the network, generally by means of capacitors, the excess of current, which flows downstream the node itself because of the reactive power demand, will be reduced upstream.

Electric utilities usually charge for a part of the reactive energy consumed. This reactive energy E_Q must be paid for, since the utility has to provide plants of sufficient size to generate and transmit both the active and reactive power required by end users and the active power corresponding to line losses.

If $E_Q = 0.5 \times E_p$ then $\cos \varphi = 0.9$.

This value, $\cos \varphi = 0.9$, represents a situation that is generally accepted by the utility companies in the delivering node. At lower $\cos \varphi$ values, corresponding to $E_Q > 0.5 \times E_p$, a charge must be paid for each kvarh exceeding 0.5 kWh; at higher $\cos \varphi$ values, corresponding to $E_q < 0.5 \times E_p$ there are no charges.

Utilities do not accept $\cos \varphi$ values exceeding unity, which may occur when capacitors remain in operation even with low load rates, because of overvoltages along the lines upstream. Consequently, every site must install control equipment or manual procedures in order to avoid this problem. The same procedures must be used to protect internal distribution lines, if capacitors are installed at the end-user nodes.

The capacitor power (kvar) needed to raise $\cos \varphi$ to a set value can be calculated with reference to the mean reactive and active power over a set period ($Q = E_Q$/operating hours, $P = E_P$/operating hours) as follows (see also Fig. 7.3):

– Reference situation:

$$Q_0, P_0, \text{ and } \cos \varphi_0 = \cos \operatorname{atan} \frac{Q_0}{P_0}.$$

– Final situation ($\cos \varphi_1 > \cos \varphi_0$)

$$Q_1, P_0, \text{ and } \cos \varphi_1 = \cos \operatorname{atan} \frac{Q_1}{P_0}.$$

$$\Delta Q = Q_0 - Q_1 = \text{power of capacitors to be installed.}$$

– If the final situation must be $\cos \varphi_1 = 0.9$

$$\cos \varphi_1 = 0.9 \text{ corresponds to } \varphi_1 = \operatorname{atan} \frac{Q_1}{P_0} = \operatorname{atan} 0.5.$$

Fig. 7.3 Power factor
control: A basic scheme
of installation

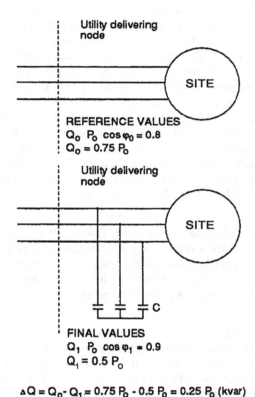

Q_o= kvarh in a given period / given hours (kvar)
P_o = kWh in a given period / given hours (kW)

Utility delivering
node

SITE

REFERENCE VALUES
Q_0 P_0 $\cos \varphi_0 = 0.8$
$Q_0 = 0.75\ P_0$

Utility delivering
node

SITE

C

FINAL VALUES
Q_1 P_0 $\cos \varphi_1 = 0.9$
$Q_1 = 0.5\ P_0$

$\Delta Q = Q_0 - Q_1 = 0.75\ P_0 - 0.5\ P_0 = 0.25\ P_0$ (kvar)

$$Q_1 = 0.5 \times P_0 \qquad \Delta Q = Q_0 - Q_1 = Q_0 - 0.5 \times P_0.$$

If the section at the utility-delivering node is considered and capacitors are
installed in this section, cost rather than energy is saved because only utility rates
for excess reactive power are reduced. Consequently, energy is saved only in utility
systems upstream of the delivering node, since these carry the same active power
with a lower current.

The foregoing considerations apply also to internal sections of power distribution systems: to save energy on the site too capacitors must be installed generally
near the loads demanding most reactive power (high-power motors, transformers,
inductive loads, etc.).

7.4   Electrical Drives

Electrical drives can be defined as systems in which electric energy is transformed into mechanical and vice versa by controlling electrical (voltage, frequency, and current) and mechanical (torque, speed) parameters. The main components of electrical drives are shown in Fig. 7.4: electrical rotating machinery, power electronic devices, and related controls with transducers of electrical and mechanical quantities (current, voltage, frequency, magnetic flux, speed, shaft position, torque, etc.).

Electrical rotating machinery may operate as a motor or as a generator depending on the power flow direction, from the shaft to the electric network or vice versa. Motors and drives transform electric energy into the mechanical energy needed at fixed speed or at variable speed to operate process and facility equipment, of which pumps, compressors, and fans are the commonest. Downstream other energy transformations are performed, e.g., from mechanical energy into cold fluid or compressed air.

A rated power or a nameplate power is introduced as the maximum output power which can be delivered as mechanical or electric power in specific conditions (voltage, current, frequency, speed, shaft profile demand and service, type of cooling) without exceeding the maximum temperature allowed for each class of insulating materials used inside the electrical machinery. These temperatures depend on the class of insulating materials, but they do not exceed 453 K (180 °C; 356 °F) for the highest class (class H in IEC standard); if operating temperatures exceed these limits, insulating materials are damaged to the extent of losing all their useful properties, thus provoking short circuits and machine failures. Attention must be paid to the cooling system (natural ventilation, forced ventilation, intermediate fluid cooling, etc.) in order to keep temperatures below the rated values.

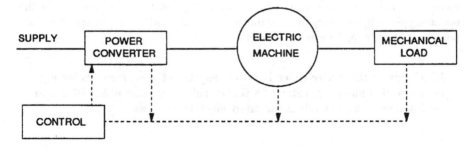

Fig. 7.4 Basic scheme of electrical drives

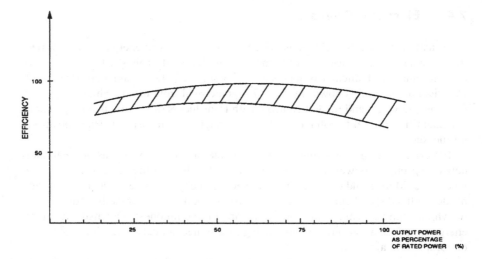

Fig. 7.5 Efficiency versus output power for electric motors

Efficiency of the rotating machinery (that is, the ratio between mechanical output power and electric input power for motor operation; the ratio between electric output power and mechanical input power for generator operation) is generally in the range of 85–95 %, the higher values for medium- and high-power machines (medium-power machines between 50 and 500 kW; high-power ones above these values). Efficiency also depends on the output power as percentage of the rated power as shown in Fig. 7.5. Smaller machines usually have lower values of efficiency.

Efficiency can be improved (1) in the design phase, with the so-called high-efficiency motors where efficiency is 2–3 % points higher than in normal-efficiency motors, by using low-loss materials, in particular low-loss iron, and (2) in the operating phase by allowing the machine to operate around the maximum efficiency point (see also Chap. 5 for transformers).

Efficiency of the entire drive is lower because of power converter efficiency, which however is about 95 %. Overall efficiencies of 85–90 % can be assumed as actual values for many electrical drives.

7.4.1 Basic Principles and Operating Modes of Electrical Machines

Electrical machines are traditionally classified as alternating current machines (synchronous machines, asynchronous or induction machines, commutator machines) and direct current machines (commutator machines) according to the voltage of the feeding line. The stationary structure (stator) and the rotary structure (rotor) are both generally provided with windings. One structure has to generate a magnetic flux by means of a particular set of windings (field windings in commutator machines and synchronous machines) or permanent magnets, and the other structure has the main windings (armature windings) through which power flows from the electric line to the shaft or vice versa. Asynchronous machines (also called induction machines) have polyphase stator windings (generally two or three phase) which generate a rotating flux and carry the electric power fed by the line; rotor windings are usually polyphase squirrel-cage windings.

A.c. (alternating current) machines can be fed directly from the line at medium or low voltage depending on the machine power (medium voltage for high-power machines, more than 500 kW). Synchronous machines must rotate at a synchronous speed related to the line frequency (synchronous speed: in r/min $= 60f$/pair of poles; in rad/s $= 2\pi f$/pair of poles where the frequency f is expressed in Hz) to effect the energy transformation, so they have to rotate at a constant speed and are not self-starting. All the other a.c. machines (except the synchronous ones) may work at any speed and are self-starting.

D.c. (direct current) machines require a d.c. voltage line that is not generally available in industrial applications. Power converters from a.c. to d.c. voltage are therefore necessary.

> As a general rule, shaft torque can be expressed as the product of magnetic flux (generated by a current flowing inside a field winding or inside a polyphase winding or by a permanent magnet in one structure) multiplied by current flowing inside the winding in the other structure (stator or rotor). This relationship is the basis of all torque control methods.

At steady state, torque-speed characteristics can be used to represent the output of the machines when operating as motors and the input when operating as generators. The working point is the intersection between the torque-speed curve of the machine and the load curve as shown in Fig. 7.6 where the characteristics of the various classes of machine are represented.

If the machine is fed directly from the line at constant voltage and frequency, there is only one torque-speed characteristic (that is, the locus of working points of the machine) and therefore only one working point is allowed for each load. Thus, at constant voltage and frequency, torque and speed values at the working point are imposed by the load and may change only if one of the load parameters varies.

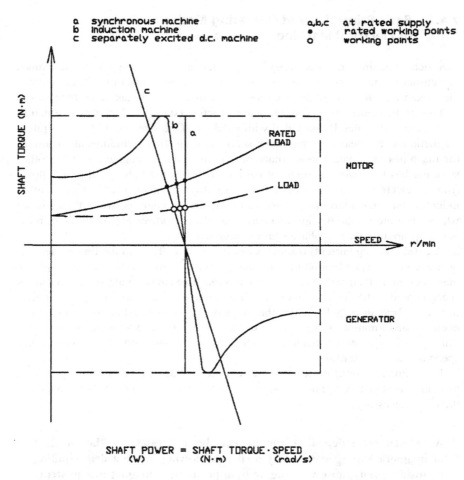

Fig. 7.6 Torque-speed characteristics for electrical machines at given input voltage and current

Rheostatic control, which changes the speed of a motor by introducing a variable amount of resistance into the stator or into the rotor circuit, always involves waste of energy when the motor runs at reduced speed. In fact, a voltage drop occurs at the terminals of the rheostat, thus provoking Joule effect losses in it. Notice that these systems cannot change the synchronous speed already defined, because they do not change the frequency of the supply. Once widely used for controlling d.c. and induction machines with wound rotors, rheostats have gradually become obsolete because of the introduction of power electronic converters which exercise a better control with lower losses.

In conclusion, energy saving at end-user level is possible only by a correct choice of the nominal power of the machine, so that the working point is set in the maximum efficiency zone, and by installing high-efficiency machines instead of standard machines. Once the installation has been made, it is the load that sets the working point, unless more complex systems, such as power electronic converters, are introduced between the supply and the electrical machine.

7.4.2   Basic Principles and Operating Modes of Electrical Drives

Unlike electrical machines directly connected to the supply, electrical drives allow a set of working points inside a working area instead of one fixed point.

The electric input variables (voltage, current, frequency) can be modified within a given range by controlling the power electronic converter properly, so that the mechanical output quantities (torque, speed, shaft position) are regulated independently within defined boundaries. In this way, the external characteristics of the machines can be continuously modified, giving rise to a family of curves which allow drive operation in an entire region of the torque-speed plane. Any point can then be reached with an appropriate control law, by varying the motor voltage (or current), the frequency, or both. The working point of the drive is no longer confined, for a given load, to the single intersection with the electrical machine characteristic resulting from a fixed-parameters source.

Two working regions are generally defined: (1) constant-torque region as a set of operating points where any torque can be delivered by the drive at any speed, within the thermal and insulation limits of the drive, and (2) constant-power region as an operating region where the torque is delivered only within the power limit of the drive (remember that shaft power = shaft torque × speed), so that its value decreases as the speed increases.

Operation at constant torque is normally performed with a constant value of the machine flux (by permanent magnets or by constant field current) and implies that any torque, within the motor's capabilities, can be delivered at any speed by action on the proper control variables, usually the current. When torque is flux multiplied by current, when the flux and the current are kept constant, a constant value of torque can be delivered at any speed by varying the voltage (and the frequency in a.c. machines), according to the control system requirements.

Operation at constant power occurs beyond the constant-torque region and higher speeds are reached at the expense of the torque within the limits of the maximum power delivered by the machine. Operation is bounded by the hyperbola of the equation power = torque × speed = constant. In this region, voltage is kept

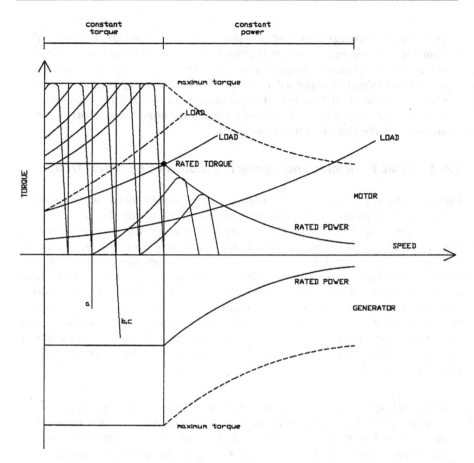

Fig. 7.7 Constant-torque and constant-power region for drives with synchronous (*a*), asynchronous (*b*), and d.c. machines (*c*)

constant (at its rated value or lower), while the machine flux is weakened. A final limit on the speed axis can be either a mechanical constraint related to the rotor type of construction or an electrical limitation (converter or motor current). Because of the constant-voltage value, constant power at any speed corresponds to a constant current from the line (neglecting losses).

Figure 7.7 shows constant-torque and constant-power regions for various drives; both motor and generator operating modes are illustrated.

Notice that electrical drives may work in a reversible mode by transforming mechanical energy into electric energy without significant modification of their structure. This capability is used mainly to effect regenerative braking by transforming the energy stored inside rotating masses at the shaft into electric energy, which is fed back to the network. Accurate braking control as well as energy recovery are the main reasons for using this mode of operation both in industry and in traction.

Electrical drives play an important role in energy saving when machines such as pumps, compressors, and fans are controlled by speed variation as an alternative to other dissipative methods (see Chaps. 10, 11, 12, 13).

In addition, electrical drives are among the main components of industrial automation systems and allow improvements in production quality and quantity, thus generally determining a decrease in specific energy consumption per unit of production. Interfacing with the plant centralized control, diagnostics and monitoring are other factors that make electrical variable-speed drives attractive for industrial applications.

7.4.3 Application Problems

Electrical variable-speed drives may be a source of thermal, mechanical, and electrical malfunctioning which must be investigated carefully in order to guarantee a safe operation of the drive itself and of the electric network.

The power factor at the drive feeding node changes according to the operating mode: generally, the lower the actual load, the lower the power factor. That depends on the internal structure of the power converter and on the current and voltage harmonics, but it can be assumed as a general rule. A control system is usually required to keep the power factor value within the accepted range.

Derating of electrical machines, when fed by power converters, is necessary because of the increase in losses due to the harmonics and because of the reduction in efficiency of cooling systems at low speed. A derating of 10–15 % depends on the harmonics; higher derating, up to 50 %, is due to thermal phenomena caused by the reduced speed range. In any case, an independent cooling system is installed, generally a fan driven by an independent electric motor.

7.5 Electrically Heated End Users

Uses of electric energy for heating include metal treatments, boilers, induction heating, electric arc furnaces, heat pumps, electromagnetic wave heating, mechanical compression of steam, etc. They are different types of applications based on different principles.

7.5.1 Joule Losses Principle

The operating principle refers to the Joule losses (power = resistance × current^2).

In metal treatment applications, two situations may occur: (1) the current is fed from the line and flows through a resistive winding; (2) the current is fed from the line and flows through an inductive winding, thus determining induced voltage and current inside the metal to be heated, which is generally inside the inductor winding.

> **Because of the losses in the power plant producing electric energy from fuels or other combustibles, electric heating does not seem convenient from the primary energy consumption point of view (roughly 6,600–10,500 kJ/kWh or 6,256–9,952 Btu/kWh are required from a utility power plant, whereas 1 kWh delivered to the end user yields 3,600 kJ or 3,412 Btu as heat).**

Heating from combustibles thus consumes less primary energy and is often cheaper, except for electric energy's special low rates (night, holiday, others) charged by utilities.

Production requirements, in terms of constant quality and quantity, may impose the use of electric heating. Notice that end-user efficiency of heating from combustibles (typical values 50–60 %) is lower than that of electric heating (the ratio between the two efficiencies is roughly 0.7), because of the unavoidable flue gas losses due to combustion, but that does not modify the considerations already made.

Efficiency can be increased by recovering heat from high-temperature flue gases to preheat combustion air (see also Chap. 15).

Boilers producing steam or hot water by using the Joule losses principle have been discussed in Sect. 6.5.

7.5.2 Electromagnetic Wave Heating

An alternative method of heating is electromagnetic wave heating. This subjects the material under treatment to electric waves ranging from a few Hz to 1 GHz in frequency for radiofrequency systems and from 1 to 300 GHz for microwave systems. Typical values are 13–27 MHz for the former and 2.5 GHz for the latter.

Advantages of these systems are the possibility of heating the material from the core to the surfaces instead of vice versa like in traditional heating systems based on fuel combustion, the reduction of local pollution, the easy control of the energy flow, and the reduction of the heating time to 1/3–1/20.

These systems are used in sterilization, defreezing, cooking, and drying. Industrial drying applications are quite common in textile, food processing, and other industries; in addition, both domestic and industrial cooking increasingly employ these systems.

Notice that these systems alone do not permit a significant energy saving in drying because the specific consumption, the energy consumed per kg of evaporated water, is roughly equal to or higher than the consumption in the case of traditional drying based on combustion.

> **Typical values are 1.2–1.3 kWh/kg of evaporated water, which are similar to those obtained when fuels are used for standard drying; of course, specific consumption as primary energy needed by utility power plants is higher than with combustion-based dryers.**

These systems are very often installed as a second step after thermal or mechanical, e.g., centrifugal, drying, to remove water from the core of the material. In this way, the quality of production is improved by action on a phase of the process in which traditional systems would require higher energy consumption to achieve similar results.

The size of single modules generally ranges from 1 to 10 kW output for microwaves and from 20 to 100 kW for radiofrequency systems. Modules are assembled for higher power.

7.5.3 Heat Pumps

A heat pump is a device that operates cyclically to transform low-temperature energy from a source (air or water) into high-temperature energy by the application of external work, mainly mechanical. The prime mover is either an electric motor or a fuel engine. Energy can be saved depending on the low-temperature source available and on the requirement of the end users at high temperature (see Sect. 12.9).

7.5.4 Mechanical Recompression and Thermocompression of Steam

This system increases the pressure and thus the temperature of saturated steam by using the mechanical energy of compression instead of thermal energy from a boiler. It is used to make small pressure changes, particularly when rejected steam from process (for instance liquid concentration or distillation) is available for reuse in the same process equipment.

Typical applications are in dairy industries, food processing, and chemical industries.

In mechanical vapor recompression systems, typical values are 20–30 kWh for recompressing 1 t of steam with a temperature increase of 6–8 K or °C (11–15 °F). Greater increases will require higher energy consumption.

For mechanical recompression in concentration plants the basic layout is shown in Fig. 7.8: typical values are 25–35 kWh of electric energy plant consumption for evaporating 1 t of water.

Lower values can be achieved if mechanical compression is combined with multiple-effect evaporation plants. If n is the number of effects, the amount of evaporated water referred to the steam consumption is $n \times 0.95^n$ kg water/kg steam.

If the compressor is driven by a turbine or a reciprocating engine with heat recovery (see cogeneration in Chap. 9), operating costs can be reduced.

An alternative to mechanical recompression is thermocompression (see Fig. 7.8) where steam pressure (and thus temperature) increases by means of an ejector (see Sect. 11.1). In this case, high-pressure steam from boilers is used to transport the rejected steam into the ejector, thus converting the velocity of the mixture to pressure in a diffuser.

The higher the driving steam pressure in proportion to the suction pressure of the rejected steam, the lower the driving steam flow rate will be.

Typical operating parameters of thermocompression are 0.6–1.2 MPa (87–174 psi) for the driving steam pressure, 0.05–0.02 MPa (7.25–2.9 psi) for the suction pressure, 1.5–2 for the compression ratio of the ejector (between sections 1 and 2 in Fig. 7.8; 15–20 K or °C or 27–36 °F of temperature increase), and 0.5–1 for the ratio between the driving steam flow rate and the rejected steam in the ejector. End-user energy consumption is ten times greater than with mechanical recompression systems.

This system can be used also in the first stage of a multiple-effect evaporation plant; in this case, the plant requires less steam or fewer effects than a pure multiple-effect evaporation plant with the same evaporation rates, but still more than plants with mechanical evaporation.

Mechanical Vapor Recompression

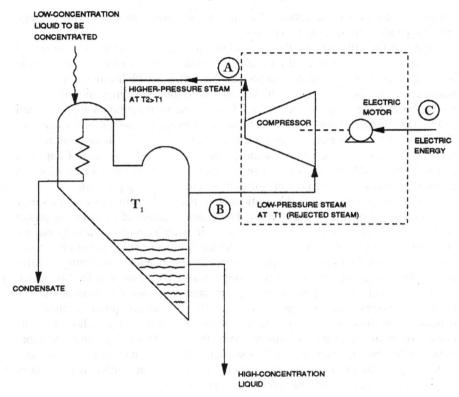

Thermal Vapor Recompression

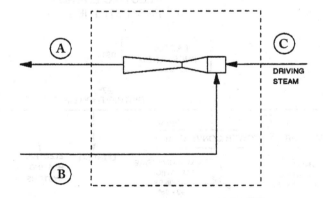

Fig. 7.8 Operating principles of MVR (mechanical vapor recompression) and TVR (thermal vapor recompression)

7.6 End Users and Load Management

Figure 7.9 shows a basic scheme for an electric system together with average efficiency values for different end users.

Note that the efficiency of the abovementioned electric end users is generally high, particularly in electrical drives, and no significant improvement can be made in the efficiency of the single component. Nevertheless, high-efficiency motors, where efficiency is 2–3 % points higher than in normal-efficiency motors, can be justified both technically and economically when the working hours are more than 4,000–5,000 h/year. Better results can be obtained by improving the performance of the whole system, for example in pumps and fans (Chap. 10), in air compressors (Chap. 11), in industrial cooling systems (Chap. 12), in HVAC systems (Chap. 13), and in lighting (Chap. 14). Of course, all these considerations can be taken as general guidelines to determine priorities in an energy-saving program.

In addition, actions such as the application of load-management techniques may give economic benefits on utility rates. This result is achieved by reducing power peaks without sacrificing production quality or quantity and by shifting energy consumption from peak hours to low-rate hours. A proper choice of the loads to be controlled and of the shutdown sequence can lead to noticeable cost saving and also to effective energy saving by reducing energy waste, particularly in facility plants.

Microcomputer or computerized energy management control systems are widely used in factories and in large buildings. They generally prove economical, depending on the complexity of the installation (number of loads, layout of the plant, power of the single controlled load); they are becoming more and more attractive for use in conjunction with non-energy functions such as monitoring and controlling pollution-producing equipment, managing preventive maintenance, recording production, and other non-engineering functions.

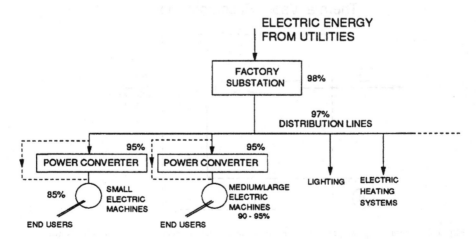

Fig. 7.9 Typical values of overall efficiency from site boundaries to electric equipment terminals

7.7   Practical Examples

Example 1 Calculate the losses in an electric industrial distribution line with different levels of power factor control.

Below is a practical example of the calculation of losses in an industrial electric distribution line with different levels of power factor control at load nodes. This approach must be followed in order to evaluate the economic validity of installing capacitors near the loads instead of placing them at the utility nodes (see Sect. 7.3).

In reference to Fig. 7.10 and on the assumption that load voltage V is kept constant, the formulae used for calculation are
- Line current

$$I = \frac{P}{\sqrt{3}\ V \cos \varphi}$$

- Total line losses

$$PL = 3 \times RT \times I^2 = \frac{RT \times P^2}{(V \times \cos \varphi)^2}$$

- Line voltage drop

$$\Delta V = \sqrt{3} \times I \times (RT \times \cos \varphi + XT \times \sin \varphi)$$

- Maximum loading power with a voltage drop of 5 %

$$PMV = \frac{5}{100} \times \frac{V^2 \times \cos \varphi}{RT \times \cos \varphi + XT \times \sin \varphi}$$

- Maximum loading power with rated current I_n

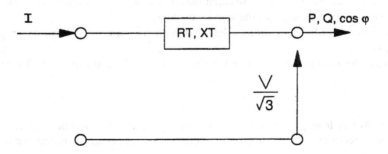

Fig. 7.10 Simplified equivalent circuit for a distribution line

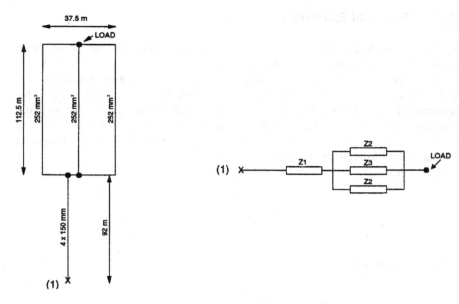

Fig. 7.11 An example of an industrial electric distribution line; see Table 7.2 for calculation

$$\text{PMI} = \sqrt{3} \times V \times I_n \times \cos\varphi$$

The two different distribution lines examined are represented in Figs. 7.11 and 7.12 together with their equivalent series circuit RT, XT.

The basic assumptions for the calculation are as follows:
– The load voltage is kept constantly equal to the nominal value.
– The recorded active and reactive power upstream the distribution line (point 1) in the site electrical substation are assumed as load powers. The load is taken to be totally concentrated at the point most unfavorable for losses and voltage drop. In a real situation the load is divided into many loads distributed along the line. The calculation used gives the maximum losses.
– Voltage drop along the line, calculated with the simplified formula as given above.

Results for the circuits illustrated in Figs. 7.11 and 7.12 are shown in Tables 7.2 and 7.3.

Case 1 Active load power and reactive load power are the reference values which have been recorded upstream the distribution line in normal operating conditions (point 1).

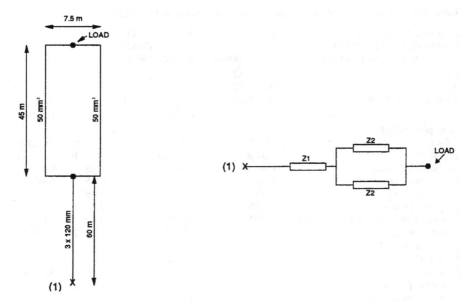

Fig. 7.12 An example of an industrial electric distribution line; see Table 7.3 for calculation

Case 2 Active load power is kept equal to the reference value as in Case 1 but a power factor equal to 0.9 is imposed because of the installation of capacitors at the load node.

Case 3 Active load power equal to the maximum loading power with power factor equal to the reference value as in Case 1 and a current corresponding to a voltage drop equal to 5 %. Alternatively, a current equal to the line nominal value is imposed if the previous value of the current was higher.

Case 4 The same as in Case 3, but a power factor equal to 0.9 is imposed because of the installation of capacitors at the load node.

Note that Case 1 represents a typical operating condition, so that calculated losses are the maximum possible in this situation. Case 2, with a power factor equal to 0.9, shows the maximum loss reduction that can be obtained with a power factor control in the operating conditions shown in Case 1.

Case 3 represents the maximum exploitation of the line with a voltage drop equal to or lower than 5 % or a line current equal to or lower than the nominal value. This case, which is different from the operating situation in Case 1, shows the maximum losses obtainable with the maximum exploitation of the line.

Case 4, with a power factor equal to 0.9, shows the maximum reduction of losses allowed by a power factor control in the operating conditions shown in Case 3.

Table 7.2 Voltage drops and losses for different cases shown in Fig. 7.11

Z1 (cable 4×150 mm^2; 92 m)		$R1 = 0.0034\,\Omega$	$X1 = 0.0014\,\Omega$		
Z2 (bus duct 252 mm^2; 150 m)		$R2 = 0.0129\,\Omega$	$X2 = 0.0196\,\Omega$		
Z3 (bus duct 252 mm^2; 112 m)		$R3 = 0.0096\,\Omega$	$X3 = 0.0146\,\Omega$		
RT		0.0073 Ω			
XT		0.0073 Ω			
Three-phase voltage		380 V, 50 Hz	P, Q in point (1)		
		Case 1	Case 2	Case 3	Case 4
Load power P, Q					
P (W)		260,000	260,000	459,200	459,200
Q (var)		300,000	125,000	529,840	220,770
$\varphi = \mathrm{atan}(Q/P)$ (rad)		0.86	0.45	0.86	0.45
Power factor $\cos \varphi$		0.65	0.90	0.65	0.90
Power losses PL (W)		7,967	4,207	24,852	13,124
Line voltage drop ΔV (V)		10.758	7.396	19.000	13.063
Energy losses working hours (h/year)		3,000	3,000	3,000	3,000
Losses (kWh/year)		23,902	12,622	74,556	39,372
Energy saving					
Case 1–Case 2 (kWh/year)			11,280		
Case 3–Case 4 (kWh/year)					35,184
Capacitor power					
Case 1–Case 2 (kvar)			175		
Case 3–Case 4 (kvar)					309

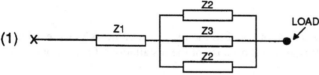

In all cases, Tables 7.2 and 7.3 shows the annual energy losses and potential energy saving if capacitors are installed as in Case 2 and Case 4. The power of the capacitors needed is also provided for these cases, so as to evaluate the investment. The cost of the capacitors alone represents a small part of the total investment which is determined mainly by installation costs dependent on many factors, such as the state of the electric line and related equipment and the distribution of loads along the line.

Economic evaluations for Case 2 and Case 4 are given in Table 19.4.

Example 2 Compare the energy consumption of a microwave and a thermal dryer system in the textile industry.

An example is given here of a microwave heating system used for drying in the textile industry.

The drying line processes 250 kg/h of dry cotton cops for 6,000 h per year. The cotton entering the line has a water content equal to 250 % by dry weight; that means 875 kg/h of total weight entering the line of which 625 kg/h is water. The water content must be reduced to 8 % by dry weight at the end of the process

Table 7.3 Voltage drops and losses for different cases shown in Fig. 7.12

Z1 (cable 3×120 mm²; 60 m)	$R1 = 0.0037\,\Omega$	$X1 = 0.0015\,\Omega$		
Z2 (bus duct 50 mm²; 52 m)	$R2 = 0.025\,\Omega$	$X2 = 0.009\,\Omega$		
RT	$0.0162\,\Omega$			
XT	$0.006\,\Omega$			
Three-phase voltage	380 V, 50 Hz	P, Q in point 1		
	Case 1	Case 2	Case 3	Case 4
Load power P, Q				
P (W)	190,000	190,000	122,160	122,160
Q (var)	265,000	90,000	171,570	57,870
$\varphi = \mathrm{atan}(Q/P)$ (rad)	0.95	0.44	0.95	0.44
Power factor $\cos\varphi$	0.58	0.90	0.58	0.90
Power losses PL (W)	11,928	4,959	4,977	2,050
Line voltage drop ΔV (V)	12.284	9.521	7.917	6.122
Energy losses working hours (h/year)	3,000	3,000	3,000	3,000
Losses (kWh/year)	35,785	14,876	14,930	6,150
Energy saving				
Case 1–Case 2 (kWh/year)		20,909		
Case 3–Case 4 (kWh/year)				8,780
Capacitor power				
Case 1–Case 2 (kvar)		175		
Case 3–Case 4 (kvar)				114

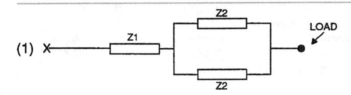

by eliminating 605 kg/h of water in two steps: (step 1) an expulsion of 500 kg of water/h by mechanical centrifugation or by compressed air in an autoclave (water content at the end of this step is 125 kg/h) and (step 2) removal of 105 kg of water/h by a thermal dryer or a microwave system.

The two systems are compared below in regard to the second step of drying.

The thermal dryer:

The dryer works with hot air produced by a steam/air exchanger and has a requirement of useful power for drying equal to 7,500 kJ/kg of expelled water. The electric consumption for ventilation is 120 kW for a rate of 105 kg of water/h.

The microwave system:

The microwave dryer removes water from the core of the material.

The rated useful power of the microwave equipment is 90 kW (30 modules of 3 kW each) with an absorbed power of 128 kW for a rate of 105 kg of water/h. The specific consumption is then $128/105 = 1.22$ kWh/kg of water.

Table 7.4 lists the main data for the comparison and the resultant energy saving. Economic evaluation is shown in Table 19.4.

Table 7.4 Comparison between microwave-based dryer and hot-air dryer

	Operating parameters		Hot air (a)	Microwave (b)
a	Working hours Drying (step 2)	h/year	6,000	6,000
b	Input dry cotton in cops	kg/h	250	250
c	Water content	kg/h	125	125
d	Water to be eliminated	kg/h	105	105
$e = 6,500 \times d/3,600$	Useful power for drying	kW	190	90
f	Absorbed electric power			
	For drying	kW	0	128
	For ventilation	kW	120	0
$g = e/0.87$	Absorbed thermal power[a]	kW	218	0
$h = g \times 3,600/41,860$		kgoil/h	18.8	
	Specific consumption			
$i = f/d$	Electric	kWh/kg H_2O	1.143	1.219
$l = h/d$	Thermal	kgoil/kg H_2O	0.179	0
$m = l \times 41,860$		kJ/kg H_2O	7,500	
	Annual energy consumption			
$n = f \times a$	Electric	kWh/year	720,000	768,000
$o = h \times a$	Thermal	kgoil/year	112,875	0
$o' = o/1,000$		TOE/year	112.875	0
	Energy saving (a) − (b)			
$p = n(a) - n(b)$	Electric	kWh/year		−48,000
$q = o'(a) - o'(b)$	Thermal	TOE/year		112.875

[a]Boiler efficiency equal to 0.87 (referred to LHV)

Example 3 For a given rated output power compare a high-efficiency with a normal-efficiency electric motor.

In many manufacturing industries (bricks, tiles, foundries, paper, etc.) and in HVAC systems, motor-driven fans and pumps work at constant load for the whole year.

A 55 kW rated power electric motor works at 80 % of its rated power for 5,000 h/year with 86 % efficiency.

This motor can be replaced by an electric motor having the same rated power but with higher efficiency (90 %).

Table 7.5 shows the calculation of the energy saved, which proves to be 11,370 kWh/year.

The economic evaluation is in Table 19.4.

Table 7.5 Comparison between two electric motors with different efficiencies (Example 3)

Standard-efficiency motor			55 kW	Rated output power	
Output		Working hours	Efficiency	Input	
% of rated capacity	kW	h/year	%	kW	kWh/year
80	44	5,000	86	51	255,814 (*a*)
High-efficiency motor			55 kW	Rated output power	
Output		Working hours	Efficiency	Input	
% of rated capacity	kW	h/year	%	kW	kWh/year
80	44	5,000	90	49	244,444 (*b*)
Energy saving (*a*) − (*b*)				kWh/year	11,370

Thermal Fluid Distribution Systems

8

8.1   Introduction

Energy entering the site as fuel is distributed to end users in different ways: pipelines downstream of the boiler plants may carry thermal energy by means of heating media such as steam, hot water, hot oil, and hot air; alternatively, pipelines may carry natural gas or oil to furnaces and ovens where fuels are burned directly.

In most industrial applications, steam is certainly the least expensive and most effective heat transfer medium, if water is available. The conversion of water into vapor absorbs a large quantity of heat per kilogram of water. The resulting steam, which is easy to distribute around the site, contains a relatively high amount of heat if compared with other heat transfer media. Finally, the process of heat transfer by condensation at end users is generally very efficient because of the quantities transferred in a relatively short time by means of simple equipment.

Cold fluid distribution lines deliver energy for plumbing, HVAC, and process refrigerating systems at low temperatures.

Natural gas is generally delivered at high pressure (1.5–5 MPa; 145–725 psi) and then brought down to lower pressure at site boundary or near the end users. Losses resulting from these transformations are relatively low.

8.2   Components of Steam and Other Fluid Systems

Figure 8.1 shows a typical arrangement for a steam system in an industrial plant.

The boiler produces steam at a pressure, and consequently at a temperature slightly higher than those required by the end users. The steam is carried from the boiler plant through steam main and secondary transfer lines to the process equipment. If some equipment requires a lower temperature, the steam will be throttled to lower pressure by means of a pressure-regulating valve or, if large quantities of steam are involved, by means of a backpressure turbine.

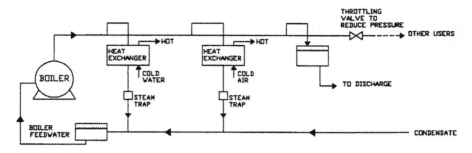

Fig. 8.1 Typical steam system circuit

In most applications, steam/water or steam/air heat exchangers are the part of the end-user equipment where heat transfer by condensation takes place.

Depending on process conditions, condensate can be discarded or recovered by steam traps, located near the steam end users, which allow condensate to drain into the condensate return line, along which it flows back to a tank and then into the boiler as feedwater.

In addition to the basic system shown in Fig. 8.1, other components are used to ensure safe operation of the system.

Insulation of pipelines, tanks and also valves and flanges and other components, represents a potential energy saving that must always be investigated to check whether it is economically valid.

The steam trap is one of the most important components of steam and condensate systems. Its main functions are to allow condensate to be drained completely from end-user equipment as quickly as possible and to facilitate the removal of air from the steam pipelines. Condensate in the steam chamber reduces heat transfer effectiveness and in addition, if liquid inclusions are trapped in a steam line, they can be accelerated to high velocity and impelled against an obstruction such as valves or other components, with an impact similar to that obtainable by hitting the component with a hammer. This phenomenon is called water hammer.

Air and some gas, generally released from the water in the boiling process, when mixed with steam make a gas mixture in which each component contributes to the pressure proportionately to its share of space available. In consequence, the pressure of the steam is reduced, as is the temperature. This means that the heat transfer is reduced or that more steam must be produced to obtain the same heating effect. In addition, the air very near the heat transfer surface will act as an insulating layer, thus further reducing heat transfer.

For steam distribution lines, temperatures generally range between 373 K (100 °C; 212 °F) and roughly 773 K (500 °C; 932 °F) for superheated steam. Higher temperatures can be found in particular applications.

Other fluid distribution lines work at different temperatures in a range from 233 K (−40 °C; −40 °F) to 373 K (100 °C; 212 °F): plumbing, HVAC, and process refrigerating systems. If oil is used as the thermal medium for industrial applications, temperatures range between 523 K (250 °C; 452 °F) and 623 K (350 °C; 662 °F).

> Typical values of velocity are 20–40 m/s for steam and 2–4 m/s for water. The volumetric flow rate can be calculated by multiplying the fluid velocity (m/s) by the pipe section (m^2). Mass flow rate is equal to volume flow rate multiplied by fluid density (kg/m^3). See Sect. 2.4 and Table 6.4 for steam density values.

8.3 Basic Principles of Pipeline Losses and Insulation

> Insulation of any hot or cold system plays an important role in energy saving. Insulation can reduce by at least 90 % the unwanted heat transfer, that is distribution losses occurring with a bare surface.

To better understand the economic criteria for the thickness of pipeline and insulation, the basic concepts of heat transfer are briefly reviewed below.

The same concepts can be applied to any problem concerning heat transfer and thermal energy balance.

Thermal energy is transferred in three main modes: conduction, radiation, and convection.

Conduction: heat is transferred through both fluids (gas and liquid) and solids when the two sides of a volume of these materials are at different temperatures. The thermal power transfer is proportional to the temperature gradient through a proportionality factor k, called thermal conductivity, and the surface.

Thermal conductivity (or k-value in insulation technology) is the measure of the capability of a material to transmit heat. Substances with high values of thermal conductivity, such as copper, are good thermal conductors, and those with low conductivity, such as polystyrene foam or cork, are good insulators. Thermal conductivity is expressed as the quantity of heat that will be conducted through unit area of a layer of material of unit thickness with unity difference of temperature between the faces in unit time. The Si unit is W/m × K; the English unit commonly used is: (Btu/h × ft × °F).

The expression of the thermal conduction power transfer Q through a surface A (Fourier's Law) is as follows:

$$Q = k \times A \times \Delta t_x \qquad \text{(W)}, \text{(Btu/h)}$$

where Δt_x is the temperature gradient between the two surfaces (K/m; °C/m; °F/ft). The sign of Q, minus in accordance with the laws of thermodynamics, is omitted here.

Table 8.1 Average values of thermal conductivity of different materials at room temperature

Material	Thermal conductivity k in the temperature range 273.15–373.15 K 0–100 °C 32–212 °F		
Pure aluminum	W/m × K	(a)	228
	Btu/h × ft × °F	$(b) = (a)/1.731$	132
	Btu × in/ft^2 × °F × h	$(c) = (b) × 12$	1,581
Cast iron	W/m × K		48
	Btu/h × ft × °F		28
	Btu × in/ft^2 × °F × h		333
Fiber glass	W/m × K		0.058
	Btu/h × ft × °F		0.034
	Btu × in/ft^2 × °F × h		0.402
Ceramic fiber	W/m × K		0.055
	Btu/h × ft × °F		0.032
	Btu × in/ft^2 × °F × h		0.381
Cellular polyurethane	W/m × K		0.025
	Btu/h × ft × °F		0.014
	Btu × in/ft^2 × °F × h		0.173

Thermal conductivities of various materials at room temperature or higher are reported in Table 8.1. If the temperature rises, thermal conductivity decreases with homogeneous metals and it increases in composite materials almost linearly in an interval range of 323–373 K (50–100 °C; 122–212 °F). With insulating materials, because of the composite or multilayer structure where also air can be present, the k coefficient must take into account the presence of all the heat transfer basic modes and its value is assessed on empirical bases. For detailed evaluations, manufacturer specifications should be consulted.

Radiation: heat is transferred as radiant energy and does not require any medium of diffusion. Radiation can take place even in a vacuum. Solid surfaces, gases, and liquids all emit, absorb, and transmit thermal radiation in varying degrees.

The rate at which energy is emitted from a system with a surface A situated inside a large space is quantified by the basic relationship:

$$Q = \varepsilon \times \sigma \times A \times (T_s^4 - T_o^4) \qquad \text{(W)}$$

where T_s and T_o are the surface and output surrounding absolute temperatures (K), ε $(0 < \varepsilon < 1)$ is an adimensional quantity that indicates how effectively the surface radiates ($\varepsilon = 1$ for the black body or the perfect emitter), and σ is the Stefan–Boltzmann constant equal to 5.67×10^{-8} W/(m^2 × K^4). As thermal radiation is associated with the fourth power of surface absolute temperature, the importance of this mode of heat transfer increases rapidly with the temperature.

Values of ε for different operating conditions (materials, different states of surfaces) are shown in Table 8.2. As with the conduction energy transfer, different units are commonly used alternatively to the SI units.

Table 8.2 Total emissivity average values (ε)

Surface	Temperature range		ε
	°C	°F	
Aluminum			
Commercial sheet	200–600	400–1,100	0.09
Heavily oxidized	100–550	200–1,000	0.3
Iron			
Steel polished	40–250	100–500	0.085
Steel oxidized	250	500	0.8
Brick			
Red	40	100	0.93
White refractory	1,100	2,000	0.29
Concrete	40	100	0.94
Glass	40	100	0.94

The abovementioned relationship is valid if some basic assumptions are respected such as transparent medium, surface with particular spectrum emission behavior, etc. For pipelines and tanks in industrial applications, these hypotheses are widely accepted and the relationship can be used with a proper choice of parameters.

Convection: heat is transferred from a solid surface at one temperature to an adjacent fluid, liquid or gas, at another temperature, when motion occurs. The thermal power transfer from the system with a flat A surface is expressed as follows:

$$Q = h \times A \times (t_s - t_f) \qquad (\text{W})$$

where t_s and t_f are the surface and fluid temperatures and h is an empirical parameter, depending on operating conditions and surface geometry, called the heat transfer coefficient. The SI unit is $W/m^2 \times K$; the English unit commonly used is: $(\text{Btu/h} \times ft^2 \times °F)$.

In most cases, heat is transferred by a combination of the abovementioned basic modes through composite or multilayer systems, and the evaluation is generally made by means of empirical parameters and linearization of the heat transfer relationships. A simplified approach based on a monodimensional scheme is generally introduced in industrial applications. As a rule, radiation is included in the heat transfer coefficient in a wide range of temperature values.

Values of h for different operating situations with surface exposed to the air are given in Table 8.3.

Values of h for fluid flowing in pipelines depend on many factors (temperature, speed, state of pipe, etc.) and must be calculated by using special formulae. They are greater than those for air and they range between hundreds and thousands $W/m^2 \times K$ (see also Chap. 15).

Table 8.3 Typical values of heat transfer coefficient (h)

			(a) W/m^2 × K (b) Btu/h × ft^2 × °F (b)=(a) × 0.1761			
Surface Emittance	Wind velocity		**Heat transfer from a hot solid surface to air (h)**			
	km/h	Operating temperature °C °F	100 212	200 392	300 572	400–550 752–1022
1.0	0	(a) (b)	9.31 1.64	11.13 1.96	12.34 2.17	13.52 2.38
	3		11.83 2.08	13.85 2.44	15.35 2.70	16.22 2.86
	6		13.85 2.44	16.70 2.94	18.32 3.23	19.58 3.45
0.6	0		7.28 1.28	9.01 1.59	9.96 1.75	10.32 1.82
	3		10.32 1.82	12.62 2.22	13.85 2.44	14.20 2.50
	6		12.62 2.22	14.94 2.63	16.22 2.86	16.70 2.94
0.2	0		5.56 0.98	7.14 1.26	7.69 1.35	8.33 1.47
	3		8.48 1.49	10.32 1.82	11.13 1.96	11.59 2.04
	6		10.32 1.82	12.91 2.27	14.20 2.50	14.94 2.63

For low temperature service, values are lower. For higher wind velocity, values are higher

	Heat transfer from liquid or gas to a solid surface (h)
From steam to a solid surface	(a) 4000–3000 W/m^2× K (b)704–528 Btu/h×ft^2×°F
From water to a solid surface	(a) 400–200 W/m^2× K (b) 70–35 Btu/ h× ft^2 × °F
From heavy organic liquid to a solid surface	(a) 100–50 W/m^2× K (b) 18–9 Btu/ h× ft^2 × °F

In a multilayer system, heat is transferred from one fluid (fluid inside i) at higher temperature (t_i) through an n-layer slab to another fluid (fluid outside o) at lower temperature (t_o).

The fundamental equation for the heat transfer calculation is:

$$\text{Heat flow through unit area} = \frac{\text{fluid temperature difference}}{\text{overall thermal resistance}} = \frac{t_i - t_o}{R_{th}}$$

$$= (t_i - t_o) \times U \left(\frac{W}{m^2}\right) \left(\frac{Btu}{h \times ft^2}\right)$$

where

t_i = inside fluid temperature (K), (°C) (°F),

t_o = outside fluid temperature (K), (°C)(°F),

R_{th} = overall thermal resistance $\left(\frac{m^2 \times K}{W}\right) = \left(\frac{m^2 \times °C}{W}\right) \left(\frac{h \times ft^2 \times °F}{Btu}\right)$

$U = 1/R_{th}$ = overall heat transfer coefficient (see Chap. 15)

Note that:

- For homogeneous materials, thermal resistance is the reciprocal of the system thermal conductance which represents the quantity of heat that will be conducted through unit area of a layer of material of a defined thickness with unity difference of temperature between the two fluids in unit time;
- For n-layer systems thermal resistance R_{th} is the reciprocal of the system thermal conductance which represents the quantity of heat conducted through unit area of a n-layer material of defined thicknesses with unity difference of temperature between the two fluids in contact with the end faces in unit time;
- The above given relationship is based on the assumption that the radiation heat transfer is negligible or that its effect is introduced into the R_{th} by linearizing the radiation law. In the same way, all heat transfer models are linearized;
- At thermal steady state, the same amount of heat flows through each layer of the system. Thus, the temperature of each layer can be calculated.

Specialized handbooks on this subject should be consulted for more detailed evaluations.

Examples are given below of flat-surface and of pipeline multilayer systems with bare surfaces and with insulation.

8.3.1 Flat-Surface Multilayer System (See Fig. 8.2)

When a flat surface is heated on one side and cooled on the other, as shown in Fig. 8.2, heat flows from the hot side to the cold side.

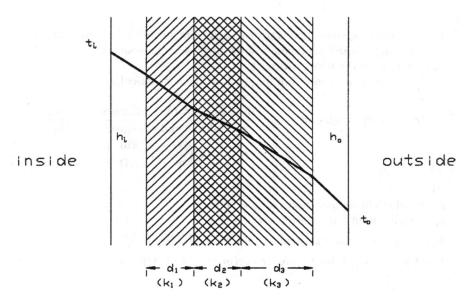

Fig. 8.2 Heat transfer through a multilayer wall

The flow of heat is defined as follows:

$$Q = A \times \frac{t_i - t_o}{R_{th}}$$

where $A =$ internal and external surfaces with the same value (m^2) (ft^2),

$$R_{th} = \frac{1}{h_i} + \sum_1^n j \frac{d_j}{k_j} + \frac{1}{h_o} \left(\frac{m^2 \times K}{W}\right) = \left(\frac{m^2 \times °C}{W}\right) \quad \left(\frac{ft^2 \times h \times °F}{Btu}\right),$$

$d_j =$ thickness of the j layer (m) (ft),
$t_i =$ inside temperature (K) (°C) (°F),
$t_o =$ outside temperature (K) (°C) (°F),
$k_j =$ thermal conductivity of the j layer $\left(\frac{W}{m \times K}\right) = \left(\frac{W}{m \times °C}\right) \quad \left(\frac{Btu}{ft \times h \times °F}\right),$
$d_j/k_j =$ thermal resistance of the j layer $\left(\frac{m^2 \times K}{W}\right) = \left(\frac{m^2 \times °C}{W}\right) \quad \left(\frac{ft^2 \times h \times °F}{Btu}\right),$
$h_i =$ heat transfer coefficient, radiation included, between the inside fluid and the
internal surface of the first layer, $\left(\frac{W}{m^2 \times K}\right) = \left(\frac{W}{m^2 \times °C}\right) \quad \left(\frac{Btu}{ft^2 \times h \times °F}\right),$
$h_o =$ heat transfer coefficient, radiation included, between the external surface of
the last layer and the fluid outside $\left(\frac{W}{m^2 \times K}\right) = \left(\frac{W}{m^2 \times °C}\right) \quad \left(\frac{Btu}{ft^2 \times h \times °F}\right)$

If one of the layers is made of insulating material, this layer has the highest
thermal resistance (d_j/k_j) and the heat flow is less than with a bare surface.

Table 8.4 shows heat losses from a composite wall. Notice that heat
transfer through the layers is reduced by insulation to roughly 10 %.

Table 8.4 Heat losses from a composite wall (reference Fig. 8.2)

Thickness of insulation (d_2)		Overall thermal resistance		Heat losses Q		Energy saving		
mm	$10^{-3} \times$ ft	$m^2 \times K/W$	$ft^2 \times h \times °F/Btu$	W/m^2	$Btu/h \times ft^2$	W/m^2	$Btu/h \times ft^2$	$kg_{oil}/h \times m^2$
0	0.0	0.190	1.079	526	167	0	0	0.000
10	32.8	0.372	2.112	269	85	257	82	0.026
20	65.6	0.554	3.144	181	57	346	110	0.035
30	98.4	0.735	4.177	136	43	390	124	0.039
40	131.2	0.917	5.209	109	35	417	132	0.042
50	164.0	1.099	6.242	91	29	435	138	0.044
60	196.9	1.281	7.274	78	25	448	142	0.045
70	229.7	1.463	8.307	68	22	458	145	0.046
80	262.5	1.645	9.340	61	19	466	148	0.047
90	295.3	1.826	10.372	55	17	472	149	0.048
100	328.1	2.008	11.405	50	16	477	151	0.048

BASIC FORMULA

$$Q = A \times \frac{t_i - t_o}{R_{th}}$$

$$R_{th} = \frac{1}{h_i} + \frac{d_1}{k_1} + \frac{d_2}{k_2} + \frac{1}{h_o}$$

INPUT DATA

wall conductivity k_1	2.500 W/m× K	1.444 Btu/h×ft × °F
fiberglass conductivity k_2	0.055 W/m×K	0.032 Btu/h×ft× °F
wall thickness d_1	0.100 m	0.328 ft
heat-transfer coefficient h_i	20 W/m²×K	3.522 Btu/h×ft² × °F
heat-transfer coefficient h_o	10 W/m² K	1.761 Btu/h×ft² × °F
t_i-t_o	100°C = 100 K	180°F

saving in $\frac{kg_{oil}}{h \times m^2} = \frac{W}{m^2} \times \frac{3.600}{\eta \times 41,860 \times 10^3}$

η combustion assumed 0.85 (LHV as reference)
LHV of oil 41,860 kJ/kg

ANNUAL ENERGY SAVING
70 mm insulation
5,000 h/year
100 m² surface
23 TOE/year

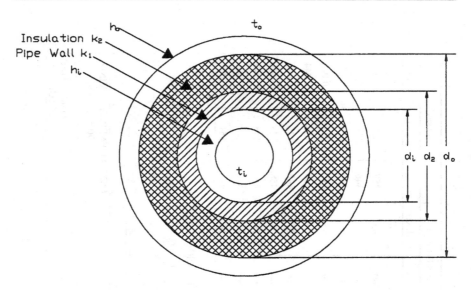

Fig. 8.3 Heat transfer through a multilayer pipe

8.3.2 Cylindrical Surface Multilayer System (See Fig. 8.3)

This example is typical of pipelines for steam or other hot fluids. The same approach can also be followed in the case of cold fluids.

The flow of heat is:

$$Q = A \times \frac{t_i - t_o}{R_{\text{th}}}$$

where $A =$ the outside ($2\pi L r_o$) or the inside ($2\pi L r_i$) surface of the pipeline of length L. Note that the R_{th} expression will be modified by the choice between the outside and the inside surface, $r_i =$ inside pipe radius, $r_o =$ outside multilayer pipe radius.

Thus:

outside surface

$$A = 2 \times \pi \times L \times r_o \text{(outside surface)} \qquad (\text{m}^2) \qquad\qquad (\text{ft}^2)$$

$$R_{\text{th}} = \frac{1}{h_i} \times \frac{r_o}{r_i} + \sum_1^n j \frac{r_o}{k_j} \times \ln \frac{r_{j+1}}{r_j} + \frac{1}{h_o} \qquad (\text{m}^2 \times^\circ \text{C/W}) \qquad (\text{ft}^2 \times \text{h} \times^\circ \text{F/Btu})$$

inside surface

$$A = 2 \times \pi \times L \times r_i \text{ (inside surface)} \qquad (\text{m}^2) \qquad\qquad (\text{ft}^2)$$

$$R_{\text{th}} = \frac{1}{h_i} + \sum_1^n j \frac{r_i}{k_j} \times \ln \frac{r_{j+1}}{r_j} + \frac{1}{h_o} \times \frac{r_i}{r_o} \qquad (\text{m}^2 \times^\circ \text{C/W}) \qquad (\text{ft}^2 \times \text{h} \times^\circ \text{F/Btu})$$

In the case of a bare tube, if the outside surface is taken as reference, the abovementioned relationships (with $n = 1$) are modified as follows:

$$Q = A \times \frac{t_i - t_o}{R_{th}}$$

where $r_2 = r_o$, $A = 2 \times \pi \times L \times r_o$

$R_{th} = \frac{1}{h_i} \times \frac{r_o}{r_i} + \frac{r_o}{k_1} \times \ln\frac{r_2}{r_1} + \frac{1}{h_o} = \frac{1}{h_i} \times \frac{r_o}{r_i} + \frac{r_o}{k_1} \times \ln\frac{r_o}{r_i} + \frac{1}{h_o}$

If the thickness $(r_o - r_i)$ of the pipe is small $(r_o - r_i < 0.1 \times r_i)$, this expression can be approximated to the following, as for a flat surface:

$$R_{th} = \frac{1}{h_i} + \frac{r_o - r_i}{k_1} + \frac{1}{h_o} \qquad \text{(simplified formula)}$$

Notice that $1/h_i$ is usually the lowest term in the case of pipelines, because of the moving fluid inside the pipe.

For insulated pipelines with one insulation layer, there follows:

$$Q = A \times \frac{t_i - t_o}{R_{th}}$$

where $A = 2 \times \pi \times L \times r_i$

$R_{th} = \frac{1}{h_i} + \frac{r_i}{k_1} \times \ln\frac{r_2}{r_1} + \frac{r_i}{k_2} \times \ln\frac{r_3}{r_2} + \frac{1}{h_o} \times \frac{r_i}{r_o}, j = 2; \; r_1 = r_i; \; r_3 = r_o$

If $r_1 = r_i$ is assumed to be equal to r_2 (small thickness of the pipe) there follows:

$$R_{th} = \frac{1}{h_i} + \frac{r_i}{k_2} \times \ln\frac{r_3}{r_2} + \frac{1}{h_o} \times \frac{r_i}{r_o}$$

If the thickness of the insulation layer is small in comparison with the bare pipe radius, that is $(r_3 - r_2) < 0.1 \times r_2 = 0.1 \times r_i$ and $(r_i/r_o) > 0.9$ the expression given above can be approximated as follows:

$$R_{th} = \frac{1}{h_i} + \frac{r_3 - r_2}{k_2} + \frac{1}{h_o} \times \frac{r_i}{r_o}$$

Table 8.5 shows heat losses from a composite metal pipeline to still air. Notice that insulation reduces heat transfer through the layers to roughly 10 %.

Attention must be paid to surface conditions such as film deposits, surface scaling, and corrosion. A fouling factor can be introduced to take account of these phenomena as an additional term in the overall thermal resistance R_{th} (see also Chap. 15).

Table 8.5 Heat losses from a composite steel pipeline to still air (see reference Fig. 8.3)

Thickness of insulation ($r_o - r_2$)	External radius (r_o)	R_{th}	Heat losses Q		R_{th} simplified	Energy saving		
mm	m	m²×K/W	W/m²	W/m	m²×K/W	W/m²	Btu/h×ft²	kg_oil/h×m²
0	0.05	0.045	3308	935	0.045	0	0	0.000
10	0.06	0.187	802	227	0.220	2506	794	0.254
20	0.07	0.308	487	138	0.396	2821	894	0.285
30	0.08	0.413	363	103	0.574	2945	933	0.298
40	0.09	0.506	296	84	0.753	3012	955	0.305
50	0.1	0.590	254	72	0.932	3054	968	0.309
60	0.11	0.666	225	64	1.112	3083	977	0.312
70	0.12	0.735	204	58	1.292	3104	984	0.314

INPUT DATA

pipe conductivity k_1	50.000 W/m×K	28.885 Btu/h×ft×°F
fiberglass conductivity k_2	0.055 W/m×K	0.032 Btu/h×ft×°F
internal radius (r_i)	0.045 m	0.1476 ft
pipe thickness	0.005 m	0.016 ft
bare pipe external radius (r_2)	0.05 m	0.164 ft
heat-transfer coefficient h_i	4,000 W/m²×K	704.4 Btu/h×ft²×°F
heat-transfer coefficient h_o	20 W/m²×K	3.522 Btu/h×ft²×°F
t_i-t_o	150°C = 150 K	270°F

saving in $\frac{kg_{oil}}{h \times m^2} = \frac{W}{m^2} \times \frac{3{,}600}{\eta \times 41{,}860 \times 10^3}$

η combustion assumed 0.85 (LHV as reference)

LHV of oil 41,860 kJ/kg

BASIC FORMULA REFERRED TO INSIDE SURFACE

$$Q = A \times \frac{t_i - t_o}{R_{th}}$$

$$R_{th} = \frac{1}{h_i} + \frac{r_i}{k_1} \times \ln\frac{r_2}{r_i} + \frac{r_i}{k_2} \times \ln\frac{r_o}{r_2} + \frac{1}{h_o} \times \frac{r_i}{r_o}$$

SIMPLIFIED FORMULA

$$R_{th} = \frac{1}{h_i} + \frac{r_o - r_2}{k_2} + \frac{1}{h_o} \times \frac{r_i}{r_o}$$

ANNUAL ENERGY SAVING
50 mm insulation
5,000 h/year
100 m² bare surface
154.5 TOE/year

As in the case of bare pipelines, the above expressions are modified if the insulated pipeline outside surface is taken as reference.

In all the foregoing relationships, average values of temperatures must be introduced in the following situations:

- When a fluid enters a pipe at one temperature (t_{in}) and leaves at another (t_{out}) an average temperature must be equal to ($t_{in} + t_{out})/2$ at any point along the pipe (see also log-mean temperature in Chap. 15);
- When the value of the parameter k for insulation material varies significantly with the temperature, the mean temperature between the two faces of the insulation must be considered. The mean value is calculated as half the sum of the temperatures on either side of the insulation.

Since the same amount of heat Q flows through each layer of a multilayer wall or pipeline, the temperature difference between the surfaces of each layer is proportional to the thermal resistance of each layer $\left(\Delta t_{Layer} = Q \times R_{th_{Layer}} \right)$.

8.3.3 Tanks, Other Equipment

In these cases specialized handbooks should be consulted to ascertain the relationships and coefficients related to the shape of the structure to insulate and to the operating temperature.

8.4 Thermal Energy End Users

At the end of thermal distribution pipelines, end users can be grouped as follows:
- Furnaces and ovens where fuels are burned directly;
- Systems where steam or hot water is released to make direct contact with the products (steam to heat a water tank, steam to sterilize food products, etc.). In this case, because of possible contamination during the process, the condensate is generally discarded; potential recovery and thus potential energy saving must be considered carefully in each individual case;
- Heat exchangers where the heat medium (steam, water, oil, etc.) transfers energy to a colder fluid (air, water, or other fluids) by reducing its temperature or by changing its state as in steam condensation (see Chap. 15);
- Thermocompression (see Sect. 7.5).

As a general rule, the steam consumed in the final users can be calculated by using a few basic relationships.

8.4.1 Heating a Mass by Steam

$$m_{\text{steam}} = (c \times m \times \Delta t)/h_{\text{v}}$$

where $m_{\text{steam}} =$ steam mass flow rate, $c =$ specific heat of the material to be heated, $m =$ mass flow rate to be heated, $\Delta t =$ temperature increase of the mass, $h_{\text{v}} =$ evaporation enthalpy or latent heat of the steam.

8.4.2 Exchanging Heat Between a Fluid and Steam (See Also Chap. 15)

$$m_{\text{steam}} = A \times \frac{\Delta t}{R_{\text{th}}} \times \frac{1}{h_{\text{v}}} = A \times \Delta t \times U/h_{\text{v}}$$

where $m_{\text{steam}} =$ steam mass flow rate, $A =$ surface of exchange, $\Delta t =$ the log-mean difference between the temperatures of the two fluids at the two sides of the exchanger (sec Chap. 15), $R_{\text{th}} =$ overall thermal resistance $= 1/U$, $U =$ overall heat transfer coefficient (see Chap. 15), $h_{\text{v}} =$ evaporation enthalpy or latent heat of the steam.

8.4.3 Extraction of Water in Industrial Dryers by Evaporation

The necessary quantity of heat medium (steam, oil) is related to the quantity of water contained in the product and to the efficiency of the system. In industrial applications at least 4,000–5,000 kJ/kg of water (1,710–2,160 Btu/lb of water) is generally required, that is roughly 1.6–2 kg of steam, to evaporate 1 kg of water contained in the product.

Operation of most process equipment can basically be derived from the above relationships; low-consumption process equipment is designed to ensure heat recovery inside the equipment itself and the minimum requirement of thermal energy (hot water, hot air, steam, hot oil).

8.5 Practical Examples

There follows examples of insulation with reference to Figs. 8.2 and 8.3.

Example 1 Compare losses from a composite wall with different thicknesses of insulation (see Fig. 8.2 and Table 8.4)

A flat surface with air on both sides is heated on one side. The difference between inside and outside fluid temperatures is 100 °C (180 °F).

The thickness of the wall is 0.1 m (0.33 ft) and the thermal conductivity is 2.5 W/m × K (1.44 Btu/h × ft × °F). Insulation with fiberglass of different thicknesses is considered.

Calculate losses and energy saving with different thicknesses of insulation.

Details of input data and results of calculation are shown in Table 8.4.

In the case of 70 mm (0.23 ft) insulation, for example:

- Losses without insulation are 526 W/m^2 (167 Btu/h × ft^2)
- Losses with 70 mm (0.23 ft) of insulation are reduced to 68 W/m^2 (22 Btu/h × ft^2)
- Reduction of losses is 458 W/m^2 (145 Btu/h × ft^2)
- If combustibles are used for heating and combustion efficiency is 85 % (Lower Heating Value as reference), the saving in input oil entering the boiler plant is $(458 \times 3,600/1,000)/(0.85 \times 41,860) = 0.046$ kg$_{oil}$/h × m^2
- If the system works 5,000 h/year the energy saved is $0.046 \times 5,000 = 230$ kg$_{oil}$/year × m^2
- With a surface of 100 m^2 (1,076 ft^2) the energy saved is $230 \times 100 = 23,000$ kg$_{oil}$/year $= 23$ t$_{oil}$/year $= 23$ TOE/year.

Example 2 Compare losses from a composite pipeline with different thicknesses of insulation (see Fig. 8.3 and Table 8.5)

A pipeline has steam inside (160 °C; 320 °F) and still air outside (10 °C; 50 °F). The difference between inside and outside fluid temperatures is 150 °C (270 °F).

The pipeline has an external radius of 0.05 m (0.16 ft); the thermal conductivity is 50 W/m × K (28.8 Btu/h × ft × °F). The steam flow rate is 2 t/h. Insulation with fiberglass of different thicknesses is considered.

Calculate losses and energy saving with different thicknesses of insulation. Details of input data and results of calculation are reported in Table 8.5.

In the case of 50 mm (0.16 ft) insulation, for example:

- Losses without insulation are 3,308 W/m^2 (1,048 Btu/h × ft^2) or 935 W/m (970 Btu/h × ft)
- With 50 mm (0.16 ft) of insulation losses are 254 W/m^2 (80 Btu/h × ft^2) or 72 W/m (75 Btu/h × ft)
- Saving is 3,054 W/m^2 (968 Btu/h × ft^2) or 863 W/m (895.5 Btu/h × ft)
- If combustibles are used for heating and if combustion efficiency is 85 % (Lower Heating Value as reference), the saving in input oil entering the boiler plant is $(3,054 \times 3,600/1,000)/(0.85 \times 41,860) = 0.309$ kg$_{oil}$/h × m^2
- If the system works 5,000 h/year saved energy amounts to $0.309 \times 5,000 = 1,545$ kg$_{oil}$/year × m^2
- With a bare pipeline surface of 100 m^2 (1,076 ft^2) (353 m pipeline or 1,158 ft length or less if it includes additional equipment insulation as equivalent surface) the energy saved is $1,545 \times 100 = 154,500$ kg$_{oil}$/year $= 154.5$ t$_{oil}$/year $= 154.5$ TOE/year.

Economic evaluations for both examples are given in Table 19.4.

Cogeneration Plants

9

9.1 Introduction

In conventional utility power plants, electric energy is generally produced with an overall efficiency, that is, the ratio between useful output and input power as fuel, in the range 35–60 %, because of the large quantity of heat discharged into the atmosphere without recovery through cooling towers, lakes, or rivers.

Differently, in cogeneration plants (heat and power plants), either useful heat and mechanical or electric power are generated from fuel, or power is produced by recovering heat from processes. An overall efficiency ranging from 60 to 85 % can be achieved, depending on the type of cogeneration plant. If useful heat is transformed into cooling media by an absorption system, the plant is called trigeneration plant.

Cogeneration is an effective method of primary energy conservation; from this point of view, cogeneration can be applied whenever it is economically justified. The correct use of this system implies a balance between electric-power requirements and process-heat requirements in quantity and in quality. If this balance does not exist, electric power must be exchanged with utility and boiler plants must produce additional heat. The discharge of unnecessary heat should be avoided; otherwise, the system works with lower efficiency like a conventional utility plant.

Before the widespread diffusion of utilities producing and distributing electric energy to end users, cogeneration plants were very common in industry.

G. Petrecca, *Energy Conversion and Management: Principles and Applications*,
DOI 10.1007/978-3-319-06560-1_9, © Springer International Publishing Switzerland 2014

Afterwards, because of local regulations and tariffications, these plants have been unevenly located and used, the choice depending on the cost of fuel and electric energy tariffs.

9.2   Forms of Cogeneration and Trigeneration

Cogeneration plants can be grouped basically into two types referred to as topping cycles and bottoming cycles.

The topping cycles produce power, mechanical or electric, before delivering thermal energy to the end users. Typical examples are the backpressure or non-condensing steam turbine, the gas turbine and combined cycles, and the reciprocating engine where exhausts are utilized as heat for end-user needs.

The heat can also be used as input into an absorption refrigerating system for cooling.

The bottoming cycle recovers thermal energy, which would normally be discarded, to produce process steam and electricity. In this cycle, first thermal energy is used for the process, and then the exhaust energy is used to produce mechanical or electric power at the bottom of the cycle. This cycle is most attractive where there is a large quantity of thermal energy at a temperature of 623 K (350 °C; 662°F) or greater associated with exothermic reactions, as in many chemical processes and in rotary kilns and furnaces. Recovery of the steam to be used in a steam turbine is the commonest bottoming cycle; for lower temperatures other cycles, like the Rankine cycle with organic fluids (ORC—Organic Rankine Cycle), are sometimes used. In this case, the electric efficiency is around 20 %, and the electric output power ranges from 200 kW to 20 MW and more.

ORC plants are also used as topping cycles, usually fed by solid biomass.

The electric generator may be synchronous or asynchronous; the choice between them depends on the working mode: if the system is independent of the utility grid, the synchronous generator must be used; if the system is interconnected with the utility grid, either type can conveniently be used. The generator voltage can be at low or medium level depending on the size and on the layout of the internal distribution network.

Table 9.1 shows the main typical parameters of cogeneration technologies based on both fossil fuel and biomass.

Figures 9.1 and 9.2 show simplified models of cogeneration and trigeneration plants fed by natural gas, which can also be used in the case of other fuels or biomasses. Electrical and heating efficiency parameters from Table 9.1 are the input together with the rated output electrical power; in trigeneration plants conversion from heating efficiency to cooling one is made by using the absorption system COP (practically less than 0.75).

Table 9.1 Technical parameters of generation power plants fed by natural gas, solid biomass, liquid biomass, and standard coal

Technology	Electrical efficiency (%)	Heat recovery efficiency (%)	Fuel	DG = distributed generation; UP = utility plant
Reciprocating engine	35–45 %	45–35 %	Natural gas	DG
Gas turbine	25–35 %	55–45 %	Natural gas	DG
Micro gas turbine	18–20 %	62–60 %	Natural gas	DG
Boiler and backpressure steam turbine	12–20 %	68–60 %	Natural gas	DG
Boiler and backpressure steam turbine	15 %	65 %	Solid Biomass	DG
Reciprocating engine	35–40 %	45–40 %	Liquid Biomass	DG
Utility plant combined cycle 800–1,000 MW	60 %	0	Natural gas	UP
Utility plant, steam condensing turbine	26 %	0	Coal	UP

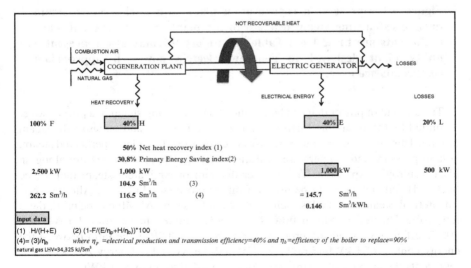

Fig. 9.1 Basic scheme of a cogeneration plant

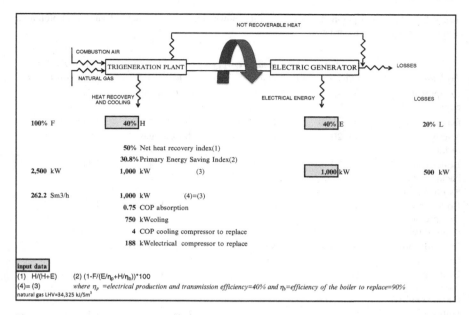

Fig. 9.2 Basic scheme of a trigeneration plant

9.3   The Backpressure or Non-condensing Steam Turbine

Figure 9.1 represents a cogeneration plant with backpressure steam turbine exhausting steam headers to the plant process. The efficiency coefficients are those from Table 9.1. Figure 9.1 may also represent a utility plant designed to generate electric power if appropriate coefficients are used.

The amount of power that can be produced by expanding steam in a prime mover is limited by the Available Energy (AE) between the inlet and outlet of the steam turbine. This energy is the enthalpy difference between the inlet superheated steam, at high pressure and temperature, and the outlet steam at lower pressure along an ideal isentropic expansion. The Mollier diagram or equivalent steam tables (see Sect. 6.4) can conveniently be used for this purpose (see Fig. 9.3). Alternatively, Theoretical Steam Rate tables such as those published by ASME can be used; these report the Theoretical Steam flow Rate (TSR) required to generate 1 kWh in a 100 % efficiency expansion process (see Table 9.2). TSR is the ratio between the energy content of 1 kWh (3,600 kJ/kWh) and the Available Energy AE (kJ/kg); it represents the amount of steam theoretically needed to produce 1 kWh:

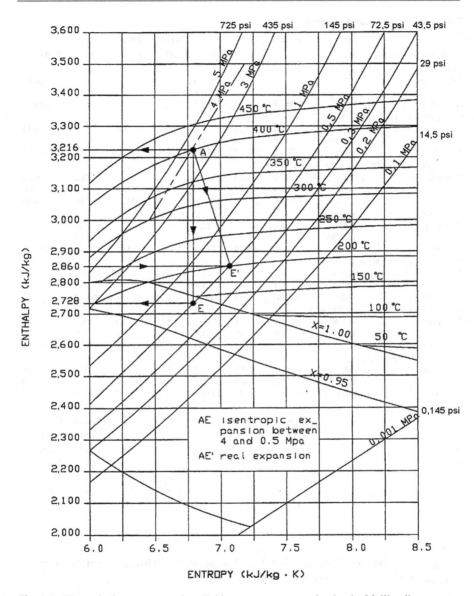

Fig. 9.3 Theoretical steam rate and available energy representation by the Mollier diagram

$$\text{TSR (kg/kWh)} = 3{,}600 \ (\text{kJ/kWh})/\text{AE (kJ/kg)}$$

where, 3,600 kJ/kWh = conversion factor (see Table 2.4) AE = $h_{\text{in}} - h_{\text{out}}$ along an isentropic expansion, h_{out} = enthalpy of the steam at outlet condition, h_{in} = enthalpy of the steam at inlet condition.

Table 9.2 Theoretical steam rate

	MPa	0.034	0.172	0.345	0.690	1.034	1.379	0.000333
Exhaust steam pressure	psi	5	25	50	100	150	200	0.048
Case 1 (Input 1.034 MPa = 150 psi)								
TSR values	lb/kWh	21.7	31.1	46.0				10.88
	kg/kWh	9.8	14.1	20.9				4.9
AE values	kJ/kg	365.74	255.19	172.53				729.475
Case 2 (Input 1.724 MPa = 250 psi)								
TSR values	lb/kWh	16.6	21.7	28.2	45.2	76.5		9.34
	kg/kWh	7.5	9.8	12.8	20.5	34.7		4.2
AE values	kJ/kg	478.11	365.74	281.44	175.59	103.74		849.752
Case 3 (Input 2.758 MPa = 400 psi)								
TSR values	lb/kWh	13.0	16.0	19.4	26.5	35.4	48.2	8.04
	kg/kWh	5.9	73	8.8	12.0	16.1	21.9	3.6
AE values	kJ/kg	610.51	496.04	409.10	299.49	224.20	164.66	987.150
Case 4 (Input 4.137 MPa = 600 psi)								
TSR values	lb/kWh	11.1	13.2	15.4	19.4	23.8	29	7.25
	kg/kWh	5.0	6.0	7.0	8.8	10.8	13.2	33
AE values	kJ/kg	715.01	601.26	515.36	409.10	333.47	273.67	1,094.715
Case 5 (Input 5.861 MPa = 850 psi)								
TSR values	lb/kWh	9.8	11.5	13.1	15.9	18.6	21.5	6.72
	kg/kWh	4.4	5.2	5.9	7.2	8.4	9.8	3.0
AE values	kJ/kg	809.86	690.14	605.85	499.16	426.70	369.14	1,181.054
Case 6 (Input 8.619 MPa = 1,250 psi)								
TSR values	lb/kWh	8.8	10.1	11.3	13.3	15.1	16.8	6.26
	kg/kWh	4.0	4.6	5.1	6.0	6.8	7.6	2.8
AE values	kJ/kg	901.89	785.81	702.36	596.74	525.60	472.42	1,267.840
Case 7 (Input 9.998 MPa = 1,450 psi)								
TSR values	lb/kWh	8.4	9.5	10.5	12.2	13.8	15.2	6.01
	kg/kWh	3.8	4.3	4.8	5.5	63	6.9	2.7
AE values	kJ/kg	944.84	835.44	755.87	659.54	575.12	522.15	1,320.579

Initial steam condition							
	Case 1	Case 2	Case 3	Case 4	Case 5	Case 6	Case 7
MPa	1.034	1.724	2.758	4.137	5.861	8.619	9.998
psi	150	250	400	600	850	1,250	1,450
°C	186	260	343	399	441	482	510
°F	366	500	650	750	825	900	950
kJ/kg	2,781	2,935	3,105	3,209	3,281	3,346	3,399
Btu/lb	1,196	1,262	1,335	1,380	1,411	1,438	1,461

Typical values of TSR are 7–8 kg of steam/kWh with a pressure drop from 4 MPa (580 psi) to 0.4 MPa (58 psi).

The previous TSR value can conveniently be converted into an Actual Steam Rate (ASR) by introducing the efficiency of the turbine, which takes into account the shift from the isentropic expansion and the efficiency of the electric generator. Then:

$$\text{ASR (kg/kWh)} = \text{TSR}/(\eta_T \times \eta_G) = (3{,}600/\text{AE}) \times 1/(\eta_T \times \eta_G)$$

Fig. 9.4 Steam turbine types for cogeneration systems: (**a**) straight non-condensing (**b**) single extraction non-condensing (**c**) double extraction condensing

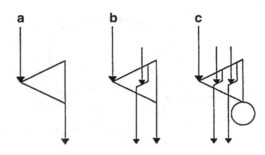

where, η_T = turbine efficiency, η_G = electric generator efficiency from the shaft to the electric output section.

The electric output is then the ratio between the Actual Steam Flow (ASF) through the turbine as required by the process and the ASR:

$$P_e \text{ (kW)} = \text{ASF (kg/h)/ASR (kg/kWh)}$$

Typical values are 10,000 kg/h of steam to produce 1,000 kW of electric power with a pressure drop of 4 MPa (580 psi).

If ASR is established, depending on the turbine cycle and operating characteristics, the relationship between the inlet and outlet steam enthalpy is as follows:

$$h_{out} = h_{in} - \text{AE} \times \eta_T = h_{in} - 3,600/(\eta_G \times \text{ASR}) \text{ (kJ/kg)}$$

Depending on the size of the turbine, on the quantity and quality of process steam demand and other operating factors, several options are available (see Fig. 9.4): straight non-condensing turbine, single or multiple extraction non-condensing turbine.

Figure 9.5 illustrates a typical industrial steam turbine cycle. It consists of a high-pressure boiler, generally 4–10 MPa (580–1,450 psi) generating superheated steam for admission to a backpressure or non-condensing steam turbine. The steam turbine drives either an electric generator or other equipment such as compressors, pumps, etc. The majority of the steam energy content remains in the outlet steam which will be utilized in the process; the energy required for mechanical power and related losses is delivered between the inlet and outlet of the turbine.

The main factors governing the optimal exploitation of a steam turbine cycle based on a fixed exhaust pressure and a constant net heat to process can be summarized as follows:

- The overall efficiency of a steam turbine plant is influenced by the inlet volume flow, inlet–outlet pressure ratio, geometry associated with turbine staging, throttling losses, mechanical coupling, and electric generator losses;

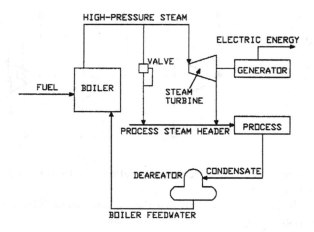

Topping Cycle

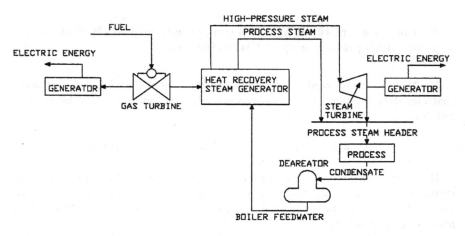

Combined Cycle

Fig. 9.5 Plant combined cycle cogeneration steam system

- Table 9.3 shows typical values of turbine and overall efficiency and ASR coefficient for a range of unit sizes of backpressure and condensing turbines. Small single-valve, single-stage units have low efficiency, less than 50 %. The multivalve, multistage units may reach an efficiency of up to 80 % in large power plants; in consequence, the greater the efficiency, the greater the electric power generated with the same steam flow. A proper sizing of the prime mover to meet process requirements is the first step in the optimization of a cogeneration plant;

Table 9.3 Overall efficiency and ASR coefficient for backpressure and condensing turbines

Type of unit	Size (MW)	AE (kJ/kg)	Range η_T (%)		Range $\eta_T \times \eta_G$ (%)		Range ASR lb/kWh		kg/kWh	
Backpressure	0.1–1	515.36	40	50	38	48	40.5	32.4	18.4	14.7
Single-valve/single-stage	1–5	515.36	65	75	62	71	24.9	21.6	11.3	9.8
Multivalve/multistage	5–25	515.36	75	80	71	76	21.6	20.3	9.8	9.2
Condensing	0.1–1	1,320.579	40	50	38	48	15.8	12.7	7.2	5.7
Single-valve/single-stage	3–20	1,320.579	70	76	67	72	9.0	8.3	4.1	3.8
Multivalve/multistage	2–50	1,320.579	76	80	72	76	8.3	7.9	3.8	3.6

Notes
Typical operating conditions for medium power backpressure turbine
See Case 4 in Table 9.2 with output pressure 0.34 MPa, 50 psi
Input steam 4.137 MPa, 600 psi
Typical operating conditions for condensing turbines
See Case 7 in Table 9.2

- The overall efficiency of a steam turbine diminishes with the output power rate, so there is always a minimum useful value below which operation is not economically worthwhile. Technical characteristics given by manufacturers must be consulted;
- An increase in the initial steam pressure and/or steam temperature will increase the amount of electric energy generated because of the increase in the enthalpy inlet–outlet difference (see Table 9.2). Since an increase of the inlet steam temperatures generally results in an increase of the temperature of the steam supplied to the process, a top limit exists to prevent outlet steam being too highly superheated, as this would necessitate a de-superheating process before delivery to the end users. In such a case, de-superheating water must be added downstream of the turbine outlet; in consequence, the steam flow through the turbine must generally be reduced by the quantity of water added, so that the electric power generated is not increased by using higher inlet steam conditions;
- An increase of the energy available for power generation is also possible if the outlet pressure is reduced with a given set of initial steam conditions. The outlet pressure must be the lowest value compatible with the end user needs;
- An increase in the power generated, with the same fuel consumption in the boiler plant, is possible if feedwater is heated by using steam extracted from turbine stages or exhausted from the process. The level of the power increase depends on the number of heaters and on the temperatures;
- To avoid damage like the erosion of the turbine blades by liquid droplets, inlet steam is always superheated. The outlet steam has generally a quality index of not less than 90 % (see quality index in Sect. 6.4).

Typical values for industrial applications with an electric power ranging between 500 and 5,000 kW are: inlet pressure 4–10 MPa (580–1,450 psi), outlet pressure 0.3–1 MPa (43.5–145 psi), ASR 10–15 kg of steam/kWh, oil consumption in the boiler plant 0.7–1 kg of oil/kWh.

An example of a backpressure steam turbine cogeneration plant is given in Sect. 9.7.

9.4   The Gas Turbine

The gas turbine as prime mover associated with a heat recovery boiler or with direct use of the exhausts in the process is another highly efficient topping cycle. It is available for both mechanical and electric generator drives in a wide range of sizes from a few hundreds to hundreds of thousands of kW. One can distinguish between industrial turbines, in the range from 1 to 400 MW, and aero-derivative turbines in the range from 2 to 40 MW. The revolution speed is generally 3,000–3,600 r/min for power higher than 60 MW and 6,000–12,000 r/min for lower power.

Gas turbine power plants may operate on either an open or a closed cycle, generally the Brayton cycle, as shown in Fig. 9.6. In the open cycle, which is the commoner, atmospheric air is continuously drawn into the compressor; then, air at high pressure enters a combustion chamber where it is mixed with fuel and combustion occurs resulting in combustion exhaust at high temperature. The combustion products expand through the turbine and are discharged into the environment. Part of the mechanical power (typically 60 %) is generally used to drive the compressor and the remainder to generate electricity or to drive other loads.

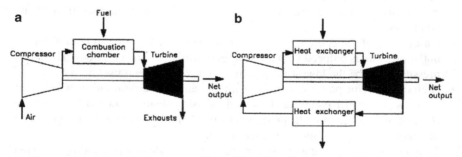

Fig. 9.6 Simple gas turbine: (**a**) open to the atmosphere; (**b**) closed

Figure 9.1 represents a cogeneration plant with a gas turbine if efficiency coefficients from Table 9.1 are used.
The maximum mechanical power available at the shaft does not generally exceed 30–32 % of the turbine input power as fuel in small and medium-sized units (1–5 MW). The amount of recoverable heat depends on the bottom temperature level required by the end user and in consequence on stack-gas temperature.
Typical input power values, in the case of natural gas input, are 350 Sm^3/h (3,330 kW) to produce 1,000 kW of electric power and 1,750 kW of recoverable heat. Specific consumptions, Sm^3 of natural gas per kWh, are lower for bigger gas turbine units.

The overall efficiency of the system, that is, the ratio of total output to input power, is roughly 70–80 %, depending on the bottom level of the thermal energy user which determines the temperature of the stack exhausts. Higher values are not generally possible, because of the great amount of airflow and so of the exhausts. Gasoil instead of natural gas can also be used with appropriate burners.

Combined cycles, where exhausts with or without supplementary burners produce steam and then electric energy in a steam turbine, can also be used for sizes higher than 20 MW of total electric output. These cycles allow for much higher power-producing capability per unit of steam than the backpressure system or the gas turbine by themselves.

Combined cycles with gas turbine and condensing or backpressure turbines are widely used with the ratio of output electric power to input power equal to 46–60 %.

The main data to take into account for selecting gas turbines are as follows:

- Unit fuel consumption/output power. Typical values range between 3 and 3.5 kW input (0.31–0.37 Sm^3/h of natural gas) per 1 kW of output shaft power. Gasoil can be used instead of natural gas;
- Exhaust flow and temperature values, on which the selection of the recovery system can be based;
- The bottom level of temperature required from process end users that limits the temperature at which the exhaust can be cooled;
- Ambient temperature on which depend the density, and so the mass of air flowing through the compressor which is a nearly constant volume flow-rate machine, as well as the bottom temperature of the cycle;
- Atmospheric pressure which varies according to the altitude. In consequence, the density and so the mass of air flowing through the compressor changes;
- Pressure drops upstream the compressor and turbine system and downstream the turbine that may influence the gas turbine useful output;

- The inlet air compression ratio and temperature, to which efficiency is closely related. Air compression ratios of 5–7 are quite common; higher ratios can be used to improve efficiency. Note that the influence of air compressor inlet temperature and of altitude is considerable since they affect the mass of air (and oxygen) available for combustion. Standard conditions for land-based systems are generally 288 K (15.6 °C; 60°F) intake at 0.1 MPa pressure (14.5 psi). A reduction in ambient air density, due to either altitude or high temperature, results in a de-rating of the turbine power output. Typical deratings are as follows: 10 % per 1,000 m of increase of altitude; 2 % per 977 Pa (100 mm H_2O; 3.9 in. H_2O) due to pressure drops downstream and upstream of the turbine; 7–10 % per 10 K or °C (18°F) of increases in ambient temperature;
- The mass of air is considerably higher than the theoretical value needed for combustion (the ratio between input combustion air and natural gas volume is 30–40) since it is imposed by the operating constraints of the air compressor-turbine system;
- If natural gas is used, the pressure of the gas which is roughly 1.5 the air pressure at the inlet of the turbine, must be at least 1–1.2 MPa (145–174 psi). In fact, typical values of air pressure are 0.6–1.5 MPa (87–217.6 psi) for industrial turbines and 1.5–3 MPa (217.6–435 psi) for aero-derivative turbines. A natural gas compressor may be required to ensure these conditions.

With steam production, heat-recovering steam generators are installed downstream of the turbine exhaust outlet. The exhaust temperature is about 723,823 K (450–550 °C; 842–1,022°F) with an excess air of 400 %. The exhaust is 15–18 % oxygen. Turbine exhausts are bypassed if they are in excess.

According to the classification made in Chap. 6, the generators may be unfired and supplementary fired.

Gas turbine exhausts can also be used directly in dryers (tiles, bricks, cereals, other productions).

Unfired generators use the exhaust of the gas turbine unit and they are convective heat exchangers.

Supplementary fired generators use a supplementary burner located downstream of the gas turbine duct to raise the temperature of the exhaust, which can be used as combustion air for it still contains a great quantity of O_2, to a maximum of 1,088–1,143 K (815–870 °C; 1,500–1,600°F). Superheated steam at high pressure is produced, which is suitable for steam turbines in combined cycle plants.

The energy recovery from the exhausts is closely related to the saturation temperature of the steam required by the process, which affects the minimum temperature value at which the exhaust can be cooled through the unfired boiler. Lower values can be obtained if an economizer is installed to pre-heat feedwater (see Sect. 6.9).

A preliminary evaluation of the amount of steam that can be generated is as follows:

$$m_{\text{steam}} = m_g \times c_p \times (t_g - t_s)/(h_s - h_i),$$

Where, m_{steam} = steam flow-rate produced by the recovery system in kg/s (lb/s), m_g = exhaust gas flow-rate in kg/s (lb/s), c_p = specific heat of the exhaust; typical value is 1 kJ/kg K (0.24 Btu/lb °F), t_g = temperature of the exhaust gas (°C, °F), t_s = saturation temperature of the steam (°C, °F), h_s = enthalpy of the superheated steam or of the saturated steam (kJ/kg, Btu/lb), h_i = enthalpy of the saturated liquid in the steam drum (kJ/kg, Btu/lb).

In the case of supplementary burners, the same relationship can be used by varying the inlet gas temperature until the desired steam flow is reached.

Typical values of operating parameters for several gas turbines are shown in Table 9.4. Notice that the output power can be regulated by varying the input fuel or by throttling the air input in the compressor.

Additional equipment to start the compressor, such as electric motor, reciprocating engine motor, or compressed air tanks, is always required.

An example of a gas turbine cogeneration plant is given in Sect. 9.7.

9.5 The Reciprocating Engine

Reciprocating engine types, principally the spark-ignited gas engine for natural gas or the Diesel engine for liquid fuel, are widely applied in cogeneration systems to drive electric generators and mechanical loads such as compressors and pumps.

Engines are available in a wide range of power, from several to thousands (kW) at different operating speeds ranging from 100 to 1,800 r/min according to the size and technical characteristics of the system.

> Figure 9.1 represents a cogeneration plant with a reciprocating engine if efficiency coefficients from Table 9.1 are used. Notice that heat recovery (50 %) is usually shared in high temperature exhaust (20 %) from which steam can be produced and hot water from cooling (30 %).

Exhaust gas, at a temperature in the range of 623–723 K (350–450 °C; 662,842°F), permits steam generation at a saturation pressure of 0.3–1 MPa, suitable for industrial applications.

Jacket and piston water cooling as well as lubricant cooling water can provide hot water at an average temperature of 343–353 K (70–80 °C; 158–176°F), which can be used for space-heating or industrial low-temperature end users. Superheated water can also be produced thanks to specially designed engine and recovery system.

> The maximum mechanical power available at the shaft generally ranges from 40 to 45 % of the engine input power as fuel. The higher values refer to turbocharged machines. The amount of recoverable heat, roughly 45–40 %, depends on the bottom temperature level required by the end user in the form of hot water.

Table 9.4 Technical parameters for standard gas turbines

| Electrical size unit | Input power | Specific consumption | | Input fuel as natural gas | | Input air flow | | | Heat recovery | Efficiency | | |
| | | | | | | | | | | Heat recovery | Electrical | Total |
MW	MW	kJ/kWh	Btu/kWh	10^6 Btu/h	Sm³/h	10^3 Sm³/h	10^3 kg/h	10^3 lb/h	MW	%	%	%
0.6	2.7	16,364	15,511	9.3	286.0	12.9	16.6	36.6	1.4	50	22	72
1	4.0	14,400	13,649	13.6	419.5	18.9	24.4	53.7	2.0	50	25	75
5	15.6	11,250	10,664	53.3	1,638.7	73.7	95.1	209.7	7.8	50	32	82
10	31.3	11,250	10,664	106.6	3,277.5	147.5	190.3	419.5	15.6	50	32	82
25	65.8	9,474	8,980	224.5	6,900.0	310.5	400.5	883.1	32.9	50	35	88
40	114.3	10,286	9,749	390.0	11,986.3	539.4	695.8	1,534.0	57.1	50	35	85
100	263.2	9,474	8,980	898.0	27,600.0	1,242.0	1,602.2	3,532.2	131.6	50	38	88
200	526.3	9,474	8,980	1,796.0	55,199.9	2,484.0	3,204.4	7,064.4	263.2	50	38	88

Notes

Air density in standard conditions 1.29 kg/Sm³ (0 °C, 32°F; 0.1 MPa, 145 psi)

Combustion air 45 Sm³ per unity of Sm³ of natural gas (standard for natural gas 15.6 °C, 60°F)

Air pressure ratio in the range 5–7

Exhaust temperature 723 K, 450 °C, 842°F; stack exhaust temperature 423 K, 150 °C, 302°F

Temperature drop available for recovery 300 K, 300 °C, 540°F

Cooling equipment must also be provided if users of hot water are not in operation. Exhausts will be discharged into the atmosphere if they are in excess.

> **Typical input-power values, in the case of natural gas input, are 250 Sm³/h (2,400 kW) to produce 1,000 kW of electric power and 1,100 kW of recoverable heat. The overall efficiency of the system, that is, the ratio between output and input power, is roughly 85 %. The ratio of output electric power to input power is roughly 40 %.**

Typical values of operating parameters for several reciprocating engines fed by natural gas are shown in Table 9.5. As a general rule, this type of prime mover is attractive if the process requires large quantities of low-level heat recovered from the jacket water and lubricant oil cooling systems which may reach half of heat rejection. An example of a reciprocating engine is reported in Sect. 9.7.

9.6 Determining the Feasibility of Cogeneration

Cogeneration feasibility is based on economic and technical factors, which have to be correlated to one another to complete a valid evaluation. Major factors for consideration are as follows:

- The ratio between electricity and fuel site demand, defined as daily, monthly, and yearly ratio. This ratio must be consistent with the ratio between electric output power and heat recovery for the chosen cogeneration system;
- The profile of heat demand, including temperature levels of end user requirements and typical fluctuations of the demand (hourly, daily, monthly, yearly). Temperature levels must be consistent with the level of heat rejected from the cogeneration system;
- The profile of electric demand and typical fluctuations as for the thermal profile. Thermal and electric profiles must be correlated with each other;
- Purchased fuel and electricity costs, present and projected future costs;
- Working hours per year and per total life of the plant;
- Plant system sized for present site needs and for the future;
- Capital cost of the cogeneration plant and operating cost during the life of the plant;
- Environmental issues.

Many cogeneration approaches can be followed in order to make a choice among system types and sizes. However, in order to ensure the highest efficiency of the system, the recovery of the rejected heat must be effective in any operating condition of the cogeneration plant. Additional boiler plants will satisfy the end user requirements, if these are higher than the recovery heat. Depending on the industrial processes, this constraint can be more or less important in determining the size of the plant.

Table 9.5 Technical parameters for reciprocating engines

Electrical size unit	Input power	Specific consumption		Input fuel as natural gas		Input air flow			Heat recovery	Efficiency		
										Heat recovery	Electrical	Total
MW	MW	kJ/kWh	Btu/kWh	10^6 Btu/h	Sm^3/h	10^3 Sm^3/h	10^3 kg/h	10^3 lb/h	MW	%	%	%
0.25	0.7	9,474	8,980	2.2	69.0	1.0	1.3	2.9	0.3	45	38	83
1	2.4	8,571	8,125	8.1	249.7	3.7	4.8	10.7	1.1	45	42	87
5	11.9	8,571	8,125	40.6	1,248.6	18.7	24.2	53.3	5.4	45	42	87
10	23.8	8,571	8,125	81.2	2,497.1	37.5	48.3	106.5	10.7	45	42	87

Notes

Air density in standard condition 1.29 kg/Sm^3 (0 °C, 32°F; 0.1 MPa, 145 psi)

Combustion air 15 Sm^3 per unity of Sm^3 of natural gas (standard for natural gas 15.6 °C, 60°F)

A first approach is designing a system which is capable of meeting thermal load requirements, regardless of electricity demand. It is connected to the utility grid and sells excess or buys additional electricity depending on the site's thermal and electric profile and on the operating conditions.

A second approach is designing a system capable of meeting either peak or base electric load requirements, regardless of the thermal demand, which nevertheless must be greater than the heat rejected. It is connected to the utility grid and sells excess or buys additional electricity depending on the sizing and on the operating conditions.

A third approach is designing a system independent of the utility grid. It requires overcapacity or redundant equipment to ensure reliability, which is guaranteed by the utility in the first two approaches. These systems have traditionally been oversized to meet peak electric demand, with supplementary equipment to satisfy the thermal demand if necessary.

9.7   Practical Examples

Three examples of cogeneration and one example of trigeneration plants are given below. A comparison of savings in primary energy is also shown.

Notice that cogeneration and trigeneration plants allow for a saving in primary energy, not in the energy entering the site. They can achieve an energy-cost saving only if the balance between electric energy and fuel energy costs is favorable for the site.

Example 1 Cogeneration plant with steam turbine

Figure 9.7 shows the energy balance and the economical evaluation of a steam turbine cogeneration plant with 1,000 kW electric power in typical working conditions (input–output pressure drop equal to 3.793 MPa, 550 psi). The heat recovery and electrical efficiency are those shown in Table 9.1. Specific consumptions due to the production of electric energy are calculated as Sm^3/kWh or kg of oil/kWh entering the steam boiler plant.

The figure shows also primary energy saving (TOE), in comparison with standard utility plants and variation of the factory energy consumption (additional fuel consumption and reduction of kWh from utilities).

The reference costs of electric energy and fuel are used for the economic evaluations. A preliminary evaluation can be made on the basis of the average cost of electric energy purchased from utilities (see Table 19.4); for a more detailed analysis, it is necessary to calculate the purchased energy and the consequent cost corresponding to the new demand profile of the plant for utilities-energy.

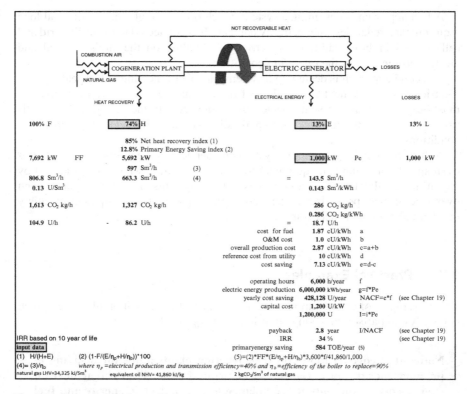

Fig. 9.7 Cogeneration plant with backpressure steam turbine

Local regulations concerning the selling of energy to the utilities and its purchase from them in emergency must also be considered.

Maintenance costs, too, must be taken into account, typically a fixed cost per unit of kWh produced.

Example 2 Cogeneration plant with gas turbine

Figure 9.8 shows the energy balance and the economical evaluation of a gas turbine cogeneration plant with 1,000 kW electric power in typical working conditions. The heat recovery and electrical efficiency are those shown in Table 9.1. Specific consumptions due to the production of electric energy are calculated as Sm^3/kWh entering the plant.

The figure shows also primary energy saving (TOE) in comparison with standard utility plants and variation of the factory energy consumption (additional fuel consumption and reduction of kWh from utilities).

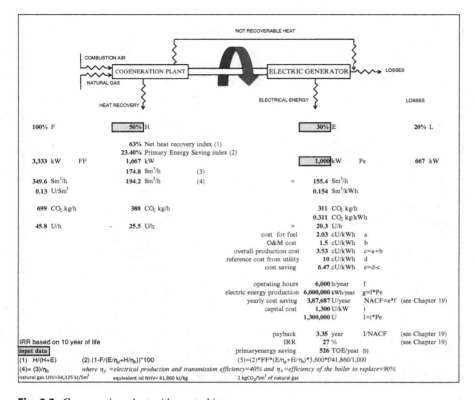

Fig. 9.8 Cogeneration plant with gas turbine

The recoverable heat form flue-gases varies according to end user requirements: hot air for drying, steam and hot water, with or without additional burners.

The reference costs of electric energy and fuel are used for the economic evaluations. A preliminary evaluation can be made on the basis of the average cost of electric energy purchased from utilities (see Table 20.3); for a more detailed analysis, it is necessary to calculate the purchased energy and the consequent cost corresponding to the new demand profile of the plant for utilities-energy. Local regulations concerning the selling of energy to the utilities and its purchase from them in emergency must also be considered.

Maintenance costs, too, must be taken into account, typically a fixed cost per unit of kWh produced.

Example 3 Cogeneration plant with reciprocating engine

Figure 9.9 shows the energy balance and the economical evaluation of a reciprocating engine cogeneration plant with 1,000 kW electric power in typical working conditions. The heat recovery and electrical efficiency are those shown in

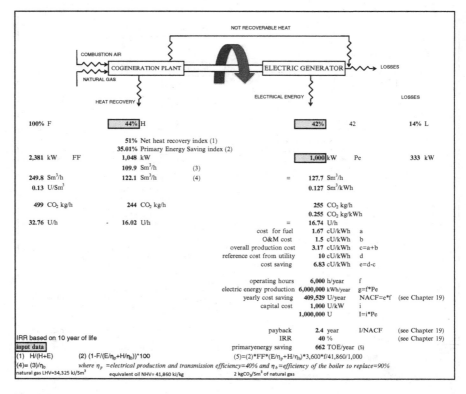

Fig. 9.9 Cogeneration plant with reciprocating engine

Table 9.1. Specific consumptions due to the production of electric energy are calculated as Sm^3/kWh entering the plant.

The figure shows also primary energy saving (TOE) in comparison with standard thermal utility plants and variation of the factory energy consumption (additional fuel consumption and reduction of kWh from utilities).

The recoverable heat from flue-gases and engine cooling media varies according to end user requirements: hot air for drying, steam and hot water at different temperatures, to which the possibility of a complete exploitation of the rejected heat is correlated.

The reference costs of electric energy and fuel are used for the economic evaluations. A preliminary evaluation can be made on the basis of the average cost of electric energy purchased from utilities (see Table 20.3); for a more detailed analysis, it is necessary to calculate the purchased energy and the consequent cost corresponding to the new demand profile of the plant for utilities-energy. Local regulations concerning the selling of energy to the utility and its purchase from them in emergency must also be considered.

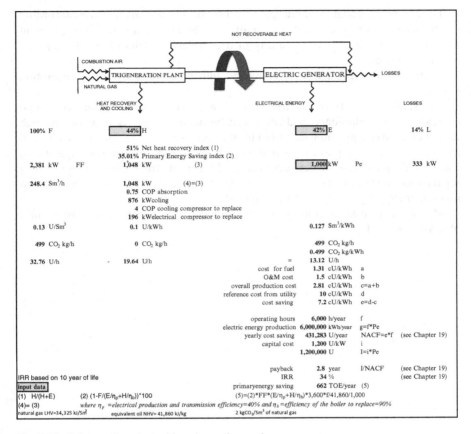

Fig. 9.10 Trigeneration plant with reciprocating engine

Maintenance costs, too, must be taken into account, typically a fixed cost per unit of kWh produced.

Example 4 Trigeneration plant with reciprocating engine

Figure 9.10 shows the energy balance and the economical evaluation of a reciprocating engine trigeneration plant with 1,000 kW electric power in a typical working condition. The heat recovery and electrical efficiency are those shown in Table 9.1. Specific consumptions attributed to the production of electric energy are calculated as Sm^3/kWh entering the plant.

The figure shows also primary energy saving (TOE) in comparison with standard thermal utility plants and variation of the site energy consumption (additional fuel consumption and reduction of kWh from utilities). Only recovered heat is taken into account for the primary energy saving (TOE) as for Example 3, regardless the subsequent conversion into cooling.

The recoverable heat from flue-gases and engine cooling media varies according to end user requirements: hot air for drying, steam and hot water at different temperatures, to which the possibility of a complete exploitation of the rejected heat is correlated.

In trigeneration plant heat is converted into cold water by using absorption systems.

The reference costs of electric energy and fuel are used for the economic evaluations. A preliminary evaluation can be made on the basis of the average cost of electric energy purchased from utilities (see Table 20.3); for a more detailed analysis, it is necessary to calculate the purchased energy and the consequent cost corresponding to the new demand profile of the plant for utilities-energy. Local regulations concerning the selling of energy to the utility and its purchase from them in emergency must also be considered.

Maintenance costs, too, must be taken into account, typically a fixed cost per unit of kWh produced.

Facilities: Pumps and Fans

10

10.1 Introduction

In process and HVAC applications, pumps and fans are widely used to move fluids, liquid (water, oil, others), or air or gas, by using mechanical energy to overcome the resistance of the flow circuit. The prime mover at the shaft is generally an electrical drive, but also other types of drive can be used. The power required at the shaft depends on the volume and the density of the fluid moved, pressure difference across the fan or the pump, and mechanical design and efficiency.

Pumps and fans can be defined as continuous rotary machines in which the rapidly rotating element accelerates the fluid passing through it, converting the velocity head into pressure, partially in the rotating element and partially in stationary diffusers or in blades.

Fans or blowers can be defined as compressors also (see Chap. 11) that develop a very low density increase from inlet to discharge (less than 5–7 % density increase).

Pumps are similar to compressors but work essentially with incompressible fluids.

For pumps moving a liquid, density variations are generally negligible. For fans moving a gas, density changes become significant and must be taken into account. Variation of density, however, is strictly related to the compression ratio; in fans with a head of 500 Pa or 0.072 psi (a quite high value for most applications), the pressure ratio equals 1.005, so the density variation is negligible.

The most commonly used machines can be classified as positive displacement or intermittent flow, centrifugal and axial machines. The characteristic curves are reported in Fig. 10.1 (see also Chap. 11 for air compressors). The basic types are shown in Fig. 10.2. Others types of machine can be used for special purposes.

In positive displacement or intermittent flow machines successive volumes of liquid are confined within a closed space and then discharged.

G. Petrecca, *Energy Conversion and Management: Principles and Applications*,
DOI 10.1007/978-3-319-06560-1_10, © Springer International Publishing Switzerland 2014

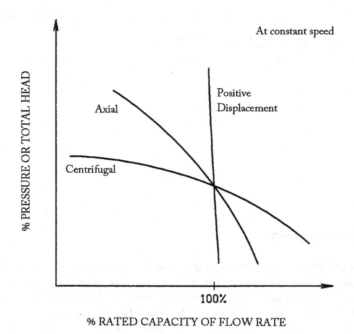

Fig. 10.1 Characteristic curves for different classes of volumetric machines

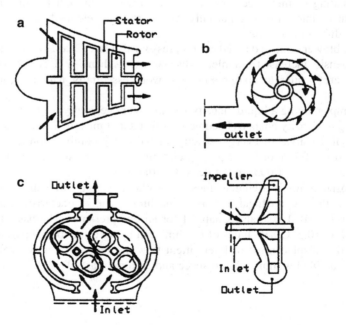

Fig. 10.2 Basic types of rotating pumps, fans, and compressors: (**a**) axial flow, (**b**) centrifugal, (**c**) lobe (intermittent flow)

Centrifugal machines are dynamic machines in which one or more rotating impellers accelerate the fluid. The main fluid flow is radial.

Axial machines are dynamic machines in which each stage consists of two rows of blades, one rotating and the other stationary. The main fluid flow is axial. The bladed rotor accelerates the fluid; the stator converts kinetic energy into pressure. Note that in low-head machines (fans, in particular), the stator does not exist or it can be unbladed.

In industrial applications, positive displacement and centrifugal machines are widely used in pumps and fans and axial machines particularly in fans.

One of the major problems in operating fans and pumps is the control of the volume flow rate to meet user needs which may require fluid flow inferior to the rated one, sometimes for long operating periods.

Common mechanical devices for the control of fans are inlet louvres, inlet vanes, outlet dampers, and different kinds of dampers such as butterfly and guillotine and variable pitch for axial fans. With these controls, an excessive power consumption occurs, together with high levels of sound and vibration and fan-volume shortfalls. Two-speed fan motors, especially if combined with either inlet-vane or inlet-louvre control, are a considerable improvement. Nevertheless, variable-speed electrical drives are the most efficient method of fluid flow control.

The most usual mechanical devices for the control of the fluid flow of pumps are throttling valves which cause additional pressure losses in the circuit and so reduce the fluid flow. Because of these losses, no significant decrease of power consumption occurs. This control, together with recycling systems, is one of the so-called dissipative methods. As with fans, the most efficient control system employs variable-speed electrical drives to regulate the shaft speed.

10.2   Basic Principles of Pump and Fan Operation

In the diagram of total pressure or total head versus volume flow rate or capacity, the working point is defined as the intersection between the characteristic curves of the machine and of the load.

In what follows, the characteristics of centrifugal pumps are discussed, but the same considerations can easily be extended to any centrifugal machines (fans, compressors) and to axial machines. The latter have a head-capacity curve much steeper than that of a centrifugal machine; hence the safe operating range is much smaller.

The basic curve for the pump at a set of given values for impeller diameter and shaft speed and the basic curve for the load, that is total pressure or head, static and friction head, versus volume flow rate or capacity, are shown in Fig. 10.3. The intersection between the two curves defines the working point $(\overline{H}, \overline{q})$ of the system, which cannot change without modification of either the load circuit or the pump.

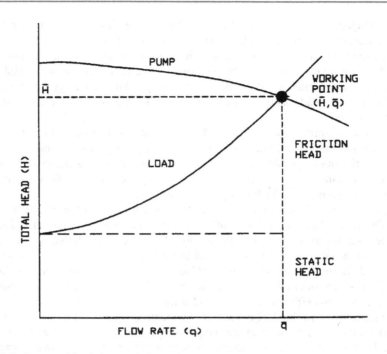

Fig. 10.3 Pump and load characteristics

The pump manufacturer provides different sets of performance curves: total head versus flow rate or capacity for different impeller diameters, shaft power versus flow rate or capacity, and sets of points with the same efficiency. Any set of curves is given for a fixed speed of the impeller (see Fig. 10.4) or for a fixed impeller diameter at different speeds.

From fluid mechanics laws and practice, the following relationships for centrifugal pumps can be derived to take into account the impeller speed variation.

If Ω_1 and Ω_2 are two impeller speeds corresponding to two working points (1, 2), and if efficiency is assumed to be constant (in practice it varies with the speed following an empirical law), it follows that:

Flow capacity $q_2 = q_1 \times (\Omega_2/\Omega_1)$ varies directly as the speed
Total head $H_2 = H_1 \times (\Omega_2/\Omega_1)^2$ varies as the square of the speed
Shaft power $P_2 = P_1 \times (\Omega_2/\Omega_1)^3$ varies as the cube of the speed

---

Curves representing flow rate and total head versus impeller speed are shown in Fig. 10.5. These relationships can be written as a function of the impeller diameter instead of the speed by simply substituting impeller diameter values D_1 and D_2 for Ω_1 and Ω_2.

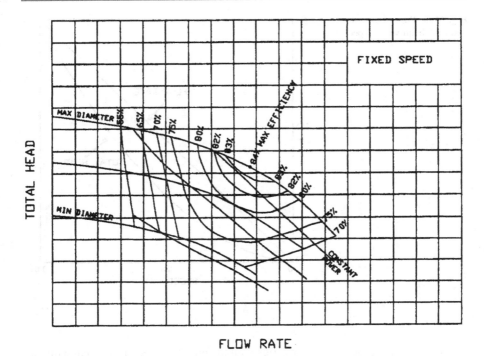

Fig. 10.4 Set of curves for a given pump

The volume flow rate can be calculated by multiplying the fluid velocity (typical values for water and air are 2–3 and 5–10 m/s) by the pipe section (m²). The mass flow rate is equal to the volume flow rate multiplied by the fluid density (kg/m³).

If other types of pumps are involved, consult the manufacturer for instructions.

From the abovementioned relationships (see also Sects. 4.7 and Table 20.1), it follows that the power P required at the shaft of pumps is

$$P = \frac{q \times H \times \rho \times (g/g_c)}{\eta} \quad (\text{W})$$

$$\frac{q \times H \times \rho \times (g/g_c) \times 0.0226}{\eta} \quad (\text{W})$$

where
q = volume flow rate (m³/s)   (gpm)
H = total head = $H_{static} + H_{losses}$ (m)   (ft)
ρ = density of the fluid (kg/m³)   (lb/gallon)
η = pump efficiency
g = gravitational acceleration 9.81 m/s²   32.17 ft/s²
g_c = conversion factor $= \dfrac{1 \text{ kg} \times (\text{m/s}^2)}{\text{N}}$ $32.17 \ \dfrac{\text{lb} \times (\text{ft/s}^2)}{\text{lbf}}$

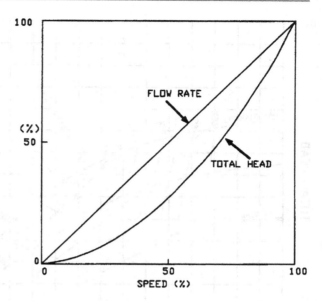

Fig. 10.5 Pump flow rate and total head versus speed

Notice that, in practice, power varies with the impeller speed more than with the cube because of the reduction of efficiency with speed.

Figure 10.6 shows two typical operating situations:

(a) Friction or dynamic lower than static losses:

Pump inlet and outlet are connected with two tanks by short pipelines without significant friction losses. If the difference in level between the tanks is constant, the static head is also constant. Friction losses, due to the movement of the fluid, are negligible as already assumed.

(b) Friction or dynamic greater than static losses:

The pipeline between the two tanks is quite long and friction losses due to the movement of the fluid must be added to the static losses.

In both situations, there is always a single working point, which is the intersection between the characteristic curves of flow capacity versus head for the pump and for the load.

From the energy-saving point of view, of particular importance is the operation at reduced flow rate in the same circuit. To reach this working point, one of the two curves, either the pump or the load curve, must be modified. As already mentioned, the pump curve varies with the shaft speed and the impeller diameter. The load curve can be modified by introducing additional equipment, such as throttling valves, between the pump and the load.

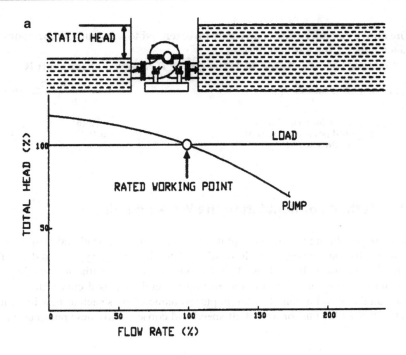

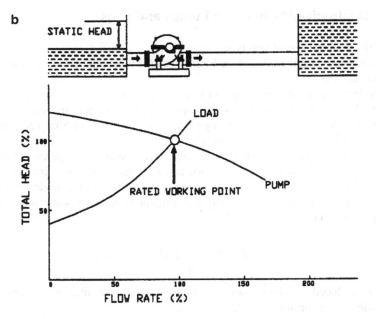

Fig. 10.6 Typical load circuits: (**a**) Dynamic losses lower than static losses; (**b**) dynamic losses greater than static losses

The above considerations can easily be extended to fans and compressors (see Chap. 11 for gas compressors).

For fans (see also Table 20.1) the power P required at the shaft is

Power shaft $P = \dfrac{q \times \Delta p}{\eta}$ (W) $\dfrac{q \times \Delta p \times 0.1175}{\eta}$ (W)

where

q = volume flow rate (m³/s)   (ft³/min)
Δp = total pressure rise (Pa)   (in H$_2$O)
η = fan efficiency

10.3 Methods of Regulating the Working Point

Methods of regulating the working point, that is, regulating head and flow, can be classed as dissipative and non-dissipative. The first category is based on the introduction of an additional head loss inside the circuit without changing the pump and fan characteristics and by shifting the end user load curve; the second is based on the introduction of different pump characteristics such as new impellers (or set of blades for fan) or new shaft speed and consequently new pump curves.

10.3.1 Dissipative Methods for Pumps and Fans

Dissipative methods are of two types:
- Introduction of head friction loss by means of a throttling valve, a damper, or similar devices
- Recycling part of the flow inside a closed-ring circuit

Figure 10.7 shows pump curves corresponding to the operating conditions in Fig. 10.6, and indicates the new working point corresponding to a 50 % reduction of the rated flow by means of a throttling valve.

If the static head is greater than the dynamic one, dissipative regulation does not provoke significant losses; in the opposite case, this regulation increases the dynamic losses to a level which would be unacceptable for a long operating period.

The power lost in the throttling valve is calculated with the same relationship used before for shaft power:

$$P_{\text{losses}} = \frac{q \times \Delta H \times \rho \times (g/g_c)}{\eta} \quad \text{(W)}$$

where q = reduced capacity value flowing through the throttling valve and ΔH = additional throttling valve head losses.

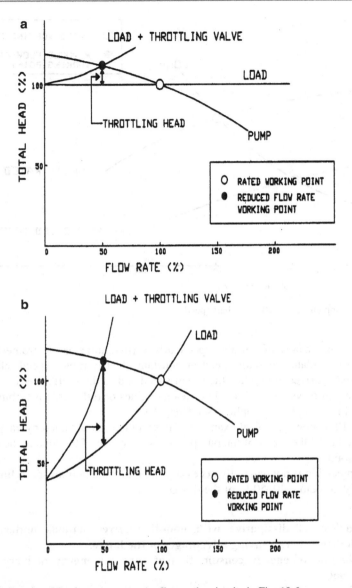

Fig. 10.7 Pump and load curves corresponding to the circuits in Fig. 10.6

10.3.2 Non-dissipative Methods for Pumps and Fans

These methods are based on modifying the pump or the fan curves without changing the load curve by (1) modifying the design of the machine (machines with variable geometry to obtain different curves, at constant shaft speed) and (2) regulating the

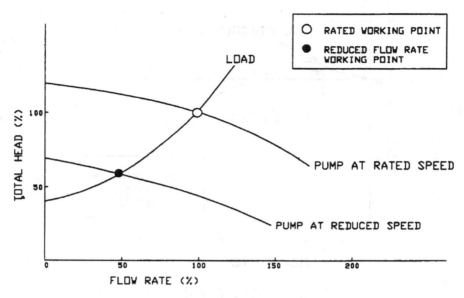

Fig. 10.8 Pump curves at different shaft speeds

shaft speed by means of variable-speed drives (hydraulic or electromechanical coupling to regulate the shaft speed at constant electric motor speed, electrical variable-speed drives to regulate the motor speed and so the shaft speed).

In large pump systems, drivers like steam engines or turbines, gas turbines, and diesel engines can also be applied (see Chap. 9).

Figure 10.8 shows a typical example of flow regulation obtained by varying the pump speed until the curve of the pump crosses the curve of the load at the desired working point.

Great precision is not generally necessary; note that the bottom speed limit does not usually go below 40–50 % of the design speed.

> **Comparison of dissipative with non-dissipative methods underlines the following energy-saving advantages of the latter:**
> • **Reduction of energy consumption at constant energy delivered to the load**
> • **Reduction of the power demanded from the electric line due to the reduction of the shaft power**

The commonest electrical drives are d.c. motors with a.c./d.c. converters and a.c. motors with inverters; the latter can also be applied to existing motors and are suggested for hazardous environmental conditions (see Sect. 7.4).

10.4 Comparison Between Dissipative and Non-dissipative Regulation Methods

An example of procedures for comparing some regulation methods is given below. The details of the calculations are set out in Sect. 10.6.

Figure 10.9 shows the situations to compare. The load curve is given together with characteristic pump curves (see Fig. 10.10), corresponding to a fixed speed (in this figure it is 3,550 r/min). The design working point is shown as $A_n (q_n; H_n; \eta_n)$.

A reduction of the design flow rate to 50 % is required. The working point should move from point A_n to A_1 ($q_1 = 0.5 \times q_n$; H_1; η_1) always along the load curve to reach the new working condition.

Two alternative procedures are considered:

- Use of a dissipative regulation that modifies the load curve at constant speed by introducing a throttling valve. The new working point is now point $A_2 \times (q_2 = 0.5 \times q_n; H_2; \eta_2)$ with the same flow as point A_1 but with increased head. This head increase corresponds to a required shaft power:

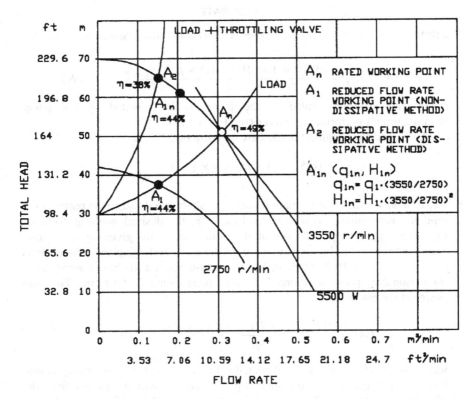

Fig. 10.9 Working points with dissipative and non-dissipative regulation (see the example in Sect. 10.6)

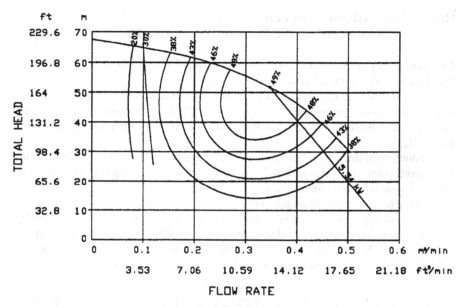

Fig. 10.10 Characteristic pump curves for a given pump at fixed speed (see the example in Sect. 10.6)

$$P \times (A_2 - A_1) = \frac{(0.5 \times q_n) \times (H_2 - H_1) \times \rho \times (g/g_c)}{\eta_2} \qquad \text{(W)}$$

where the efficiency value in the point A_2 is provided by the pump manufacturer's performance curves (see Fig. 10.10).

The power now required in the new working point A_2 is

$$PA_2 = \frac{(0.5 \times q_n) \times H_2 \times \rho \times (g/g_c)}{\eta_2} \qquad \text{(W)}$$

• Use of a non-dissipative regulation varies the shaft speed and so the pump curve until the new curve crosses the load curve at the desired working point A_1 ($q_1 = 0.5 \times q_n$; H_1; η_1). Curves at different speeds are given by the pump manufacturer or can be constructed point by point following the relationships described in Sect. 10.2, where efficiency is assumed not to change appreciably for reasonable speed variations. The new speed equals 2,750 r/min. The total power at the pump shaft is

$$PA_1 = \frac{(0.5 \times q_n) \times H_1 \times \rho \times (g/g_c)}{\eta_1} \qquad \text{(W)}$$

where the efficiency must be calculated from data supplied by the pump manufacturer. The efficiency corresponding to point A_1 will be greater than the efficiency at point A_2 (the value of the efficiency at A_1 can be assumed to be equal to that at the working point A_{1n} on the curve of nominal speed with q and H which are related to q_1 and H_1 as shown in Sect. 10.2).

- The head difference between A_2 and A_1 represents the saving in power and so in energy due to the use of a non-dissipative method. This saving depends both on the head increase and on the difference of efficiency between the two points.
- The energy saved depends on the working hours:

$$ES = (PA_2 - PA_1) \times \text{working hours} \quad (\text{Wh})$$

Typical energy saving for a 50 % reduction of the capacity ranges from 25 to 50 % of the corresponding energy needed with non-dissipative methods.

10.5 Pump and Fan Regulation Methods

Non-dissipative regulation of pumps and fans in industrial applications is an effective area of energy saving in many processes like paper, chemical, and refinery, in water-pumping systems, and in HVAC systems.

In the low- and medium-power range, from 1 to 200 kW, electrical variable-speed drives can conveniently be applied. Synchronous and asynchronous motors, fed at variable frequency and voltage, are particularly suitable for operating in hazardous locations with a proper choice of the motor and related equipment. Higher power ranges, too, can be covered with the same drives, but also gas turbines, steam motors and turbines, and diesel engines can be employed economically if the waste heat is recovered as already shown in cogeneration systems (see Chap. 9).

The amount of energy saved depends on the need for low-rate capacity, working hours of the system, and electric energy cost, but it is always significant.

Other advantages, in addition to the energy saving which is the main factor, are greater simplicity due to the elimination of throttling valves, dampers and similar devices, and simpler design of the mechanical and electric circuits made possible by softer starting.

10.6 Practical Examples

Example Compare the energy consumption of a centrifugal pump if flow is reduced to 50 % of the rated value with (1) a dissipative method and (2) a non-dissipative method.

An example is given below of technical and economic comparison between a dissipative system (throttling valve) and a non-dissipative one (electrical variable-speed drive). This example refers to a low-power pump which works at 50 % of rated capacity for all the year, but it can be assumed as a basic model for other situations.

Table 10.1(a) and (b) shows the technical characteristics of the pump and the data elaboration procedures with both SI and non-SI units. The characteristic curves

Table 10.1 Comparison between dissipative and non-dissipative methods (see Figs. 10.9 and 10.10)

Formula	Description	SI units	At rated capacity (100%) Working point		At reduced capacity (50%) Dissipative method (I) Working point		Non-dissipative method (II) Working point	
(a)								
a	Capacity	m³/min	A_n	0.312	A_1, A_2	0.16	A_1	0.16
$d = a/60$		m³/s	A_n	0.005		0.0027		0.0027
b	Head (basic load)	m	A_n	53.20	A_1	37.50	A_1	37.50
c		m			A_2	67.00		
d	Pump efficiency	%	A_n	49.00	A_2	38.00	$A_1 (A_m)$	44.00
$e = \dfrac{d \times b \times 9.81 \times r}{1{,}000}$	Power delivered to the end user circuit	kW	A_n	2.72	A_1	0.98	A_1	0.98
$f = \dfrac{d \times (c-b) \times 9.81 \times r}{1{,}000}$	Regulating losses Valve losses at pump shaft	kW		0	$A_2 - A_1$	0.77		0
	Valve losses at pump shaft	kW		0	$A_2 - A_1$	2.03		0
$g = f \times 100/d$	Total shaft pump power	kW	A_n	5.54	A_2	4.61	A_1	2.23
$h = g + e \times 100/d$	Electric motor efficiency	%	A_n	90.00		90.00	A_1	90.00
i	Electric motor losses	kW	A_n	0.62	A_2	0.51	A_1	0.25
$l = h \times (100/i - 1)$	Drive control losses	kW	A_n	0.00		0.00	A_1	0.28
m	Electric power	kW						
$n = m + 1 + h$	Supplied by the line	kW	A_n	6.16	A_2	5.12	A_1	2.75
o	Working hours	h/year	A_n	8.640	A_2	8.640	A_1	8.640
	Electric energy	kWh/year						
$p = n \times o$	Supplied by the line	kWh/year	A_n	53,222	A_2	44,237	A_1	23,760
	Energy saving at reduced capacity							
$q = p(I) - p(II)$	Method (I) – Method (II)	kWh/year					A_1	20,477
r	Density of fluid	kg/m³		1,000		1,000		1,000

(b)

Non SI units		Units	At rated capacity (100 %) Working point		Dissipative method (I) Working point		Non-dissipative method (II) Working point	
a	Capacity	ft³/min	A_n	11.02	A_1, A_2	5.651	A_1	5.651
$d' = a/60$		gpm		82.42		42.26		
b	Head (basic load)	ft	A_n	174.54	A_1	123.26	A_1	123.26
c		ft			A_2	219.82		
d	Pump efficiency	%	A_n	49.00	A_2	38.02	$A_1 (A_{1n})$	44.00
	Power delivered							
$e = a \times b \times r/60$	To the end user circuit	ft × lbf/s	A_n	2,001	A_1	725	A_1	725
$e' = e \times 1.356/1{,}000$		kW		2.7		0.98		0.98
$e'' = d' \times b \times r' \times 0.0226$		kW		2.7		0.98		0.98
$f = \dfrac{a \times (c-b) \times r}{60 \times d}$	Regulating losses Valve losses at pump shaft	ft × lbf/s		0	$A_2 - A_1$	1,490		0
$g = f \times 1.356/1{,}000$		kW				2.03		
$g' = d' \times (c-b) \times r' \times 0.0226/d$		kW				2.03		
$h = f + e \times 100/d$	Total shaft pump power	ft × lbf/s	A_n	4,080	A_2	3,400	A_1	1,647.16
$h' = h \times 1.356/1{,}000 = g' + e' \times 100/d$		kW		5.54		4.61		2.23
i	Electric motor efficiency	%		90.00		90.00		90.00
$l = h' \times (100/i - 1)$	Electric motor losses	kW	A_n	0.62	A_2	0.51	A_1	0.25
m	Drive control losses	kW		0.00		0.00		0.28
	Electric power							
$n = m + 1 + h'$	Supplied by the line	kW	A_n	6.16	A_2	5.12	A_1	2.75
o	Working hours	h/year	A_n	8,640		8,640		8,640
	Electric energy							
$p = n \times o$	Supplied by the line	kWh/year	A_n	53,222	A_2	44,237	A_1	23,760
	Energy saving at reduced capacity							
$q = p(\text{I}) - p(\text{II})$	Method (I) – Method (II)	kWh/year				20,477		
r	Density of fluid	lb/ft³		62.43		62.43		62.43
r'		lb/gallon		8.35		8.35		8.35

of the pump and of the circuit are those reported in Figs. 10.9 and 10.10, where the comparison between dissipative and non-dissipative methods is illustrated. The comprehension of correlations between the figures and the table content is facilitated by introducing working point indexes (A_n, A_1, A_2).

The energy saving is shown in Table 10.1(a) and (b), and the economic evaluation in Table 19.4.

Facilities: Gas Compressors 11

11.1 Introduction

Gas compressors are mainly used in chemical and petrochemical industries; air compressors are widely used in every kind of industries for different purposes. Air compressor applications may be broadly classified as follows: (1) providing power, as in tools or other machines and in control systems for HVAC and for industrial equipment; (2) handling process gas, usually in refineries and chemical plants, and transporting and distributing gas in pipelines; and (3) providing air for combustion as in gas turbine cycles. An additional use is for cleaning or drying material more efficiently than thermal energy.

The characteristic curves already reported in Fig. 10.1 can be used also for the different compressor types: positive-displacement compressors show a constant volume and a variable pressure ratio; centrifugal compressors (low-speed type) work with almost constant pressure ratio and variable volume; and axial and high-speed centrifugal compressors have characteristics in between those of these two types.

> **Gas compressors can be classified as (1) positive-displacement or intermittent flow, and (2) continuous flow.**

In class (1), successive volumes of gas are confined within a closed space and elevated to a higher pressure. Examples are reciprocating compressors and rotary compressors such as sliding vane, straight lobe, liquid piston, and helical lobe or screw compressors.

Class (2) comprises dynamic compressors such as centrifugal and axial compressors (see also Figs. 10.1 and 10.2) where the velocity of the rotating elements is converted into pressure partially in the rotating element and partially in stationary diffusers or blades, and also mixed-flow compressors or ejectors. Note

G. Petrecca, *Energy Conversion and Management: Principles and Applications,*
DOI 10.1007/978-3-319-06560-1_11, © Springer International Publishing Switzerland 2014

that ejector types use a high-velocity steam or gas to transport the inflowing gas; then convert the velocity of the mixture to pressure in a diffuser (see Sect. 7.5).

If the discharge pressure is kept constant, the intake pressure decreases according to the compression rate (vacuum compressor).

All major types of compressor (reciprocating, vane, screw, dynamic) are used for air compression. If the compression rate, that is the ratio between the absolute discharge pressure and the absolute intake pressure, is too high, it may become necessary to combine groups of elements in series to form a multistage system with two or more steps of compression. The gas is often cooled between stages to reduce the temperature and thus the volume entering the following stage. Theoretical minimum power with perfect intercooling and no pressure losses between stages is obtained if the ratio of compression is the same in all stages.

In what follows, the basic principles of compression processes are reported and applications to compressed-air system are pointed out.

11.2 Basic Principles of Compressed-Air Systems

The two basic theoretical compression processes, the isothermal and the adiabatic, concern both positive-displacement and continuous-flow compressors, although in practice neither of the two processes can be implemented.

11.2.1 Isothermal Compression

The temperature is kept constant as the pressure increases. This requires heat to be removed continuously and expelled from the compressor.

> **Compression from state 1 to state 2 follows the formula:**
>
> $$p_1 \times v_1 = p_2 \times v_2 = \text{constant}$$

11.2.2 Adiabatic Compression

No heat is added to or removed from the compressor during the compression isentropic process (or reversible).

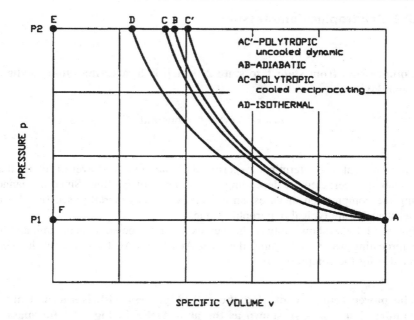

Fig. 11.1 Pressure-specific volume curves for various gas compression processes

Compression from state 1 to state 2 follows the formula:

$$p_1 \times v_1{}^k = p_2 \times v_2{}^k = \text{constant}$$

where k is the ratio of the specific heat parameters c_p/c_v, ($k = 1.4$ for air).

Figure 11.1 shows the relationship between p pressure and v specific volume for the foregoing processes; note that area ADEF represents the work required during isothermal compression; the area ABEF the work for adiabatic compression. Compressors are often designed to remove as much heat as possible but the heat of compression can never be removed as rapidly as it is generated and the temperature increases during compression. It is evident that the isothermal area is much smaller than the adiabatic; in practice, neither the isothermal nor the adiabatic compression is obtainable.

Most positive-displacement units approach an adiabatic process quite closely; dynamic units generally follow a polytrophic process with a formula similar to the adiabatic one, but the exponent is different. Of course, in all cases the exponent depends on the machine and on its operating conditions.

11.2.3 Polytrophic Compression

Compression from state 1 to state 2, which is irreversible, follows the formula:

$$p_1 \times v_1{}^n = p_2 \times v_2{}^n = \text{constant}$$

The exponent n is determined experimentally and may be lower or higher than the adiabatic coefficient k, depending on the level of cooling. Since n changes during the compression of gases, an average value is generally used ($n = 1.3$ or higher than 1.4 are typical polytrophic values).

Figure 11.1 also shows polytrophic curves for an uncooled dynamic unit and for a reciprocating water-cooled unit; the areas ACEF and AC'EF represent the work demanded for the compression.

The power requirement of a theoretical process with isothermal air compression ($T_2 = T_1$), shown as the area ADEF in Fig. 11.1 for mass unit, is as follows (notice that the relationship is transformed from the mass unit reference to the mass flow by multiplying the specific volume by the mass flow rate):

$$P_{is} = p_1 \times q_1 \times \ln\frac{p_2}{p_1} \times \frac{1}{1,000} = 101,325 \times q_1 \times \frac{1}{1,000}\ln\frac{p_2}{p_1} = 101.325 \times q_1 \times \ln\frac{p_2}{p_1} \,(\text{kW})$$

where (for English unit see Table 20.1) p_2 = outlet absolute pressure (Pa), p_1 = inlet absolute pressure (usually the atmospheric pressure = 101,325 Pa), q_1 = inlet airflow rate (m^3/s) referred to inlet condition = specific volume × inlet air mass flow rate (usually inlet conditions are standard atmospheric conditions: 0.1013 MPa; 15.6 °C, 60 °F. See footnote to Table 2.5).

The power requirement of the theoretical adiabatic compression process, shown as the area ABEF in Fig. 11.1 for mass unit, is as follows:

$$P_a = p_1 \times \frac{k}{k-1} \times \left(\left(\frac{p_2}{p_1}\right)^{\left(\frac{k-1}{k}\right)} - 1 \right) \times q_1 \times \frac{1}{1,000}$$

$$= 101.325 \times \frac{k}{k-1} \times \left(\left(\frac{p_2}{p_1}\right)^{\left(\frac{k-1}{k}\right)} - 1 \right) \times q_1 \,(\text{kW})$$

Fig. 11.2 Energy balance in a controlled-volume system

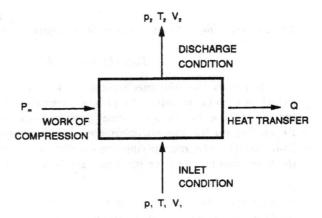

The relationship between the inlet (T_1) and outlet (T_2) temperatures is:

$$T_2/T_1 = (p_2/p_1)^{((k-1)/k)}$$

Then:

$$T_2 = T_1 \times (p_2/p_1)^{((k-1)/k)}$$

Note that all pressures and temperatures are absolute. If the compression ratio is defined, the outlet temperature depends only on the inlet temperature.

All the foregoing relationships can be used for polytrophic compression processes by substituting n for the parameter k.

Depending on the type of compression process, different heat and energy balances can be made.

Energy balance can be written with reference to a controlled-volume system as represented in Fig. 11.2:

$$
\begin{aligned}
P_m &= +Q - m \times \left(u_1 + p_1 v_1 + V_1{}^2/2 + H_1 g/g_c\right) + m \times \left(u_2 + p_2 v_2 + V_2{}^2/2 + H_2 g/g_c\right) \\
&= +Q - m \left(h_1 + V_1{}^2/2 + H_1 g/g_c\right) + m \left(h_2 + V_2{}^2/2 + H_2 g/g_c\right)
\end{aligned}
$$

where P_m = mechanical power (positive if entering the system), Q = heat transfer rate (positive if leaving the system), m = mass flow rate, h_1, h_2 = inlet, outlet enthalpy (which is defined as the sum of the internal energy, generally called u, and the product of pressure and specific volume = $u + pv$), V_1, V_2 = inlet, outlet velocity, H_1, H_2 = inlet, outlet static head, g = gravitational acceleration 9.81 m/s² = 32.17 ft/s², g_c = conversion factor = $\dfrac{1\ \text{kg} \times \left(\text{m/s}^2\right)}{\text{N}}$ $32.17\ \dfrac{\text{lb} \times \left(\text{ft/s}^2\right)}{\text{lbf}}$

If the change in static energy is neglected, as it typically happens, then:

$$P_m = Q + m \times (h_2 - h_i) + m \times \left(V_2{}^2 - V_1{}^2\right)/2$$

The contribution of kinetic energy is negligible in most cases, so:

$$P_m = Q + m \times (h_2 - h_1)$$

This means that the mechanical work equals the heat transferred (which is zero in the adiabatic process) plus the air enthalpy difference (which is zero in the isothermal process) between outlet and inlet. Depending on the known parameters, air enthalpy can be calculated from charts or by multiplying the specific heat at constant pressure by the temperature, as for an ideal gas (see Sect. 13.2).

Notice that the difference between the work required in an actual process and in an adiabatic process represents the amount of losses inside the machine.

The sum of input energy to the compressor in the form of mechanical power and any other energy as heat added or removed is transformed into the following terms: heat radiated from the frame and other parts to the atmosphere; heat conducted from the frame to the foundation; heat carried from inside by lubricating and cooling oil and radiated to the atmosphere or transferred to cooling fluid for disposal; heat carried away by the cooling medium in cylinder jackets (for reciprocating compressors); heat carried away by the cooling medium in intercoolers; and heat carried out from the compressor in the compressed air leaving the system.

The efficiency of the compression process can be defined as the ratio between the power requirement in adiabatic or isentropic compression P_{ad} and the actual power P_{actual}, at the shaft of the compressor:

$$\eta_{ad} = P_{ad}/P_{actual} = (h_{2ad} - h_1)/(h_2 - h_1) = (T_{2ad} - T_1)/(T_2 - T_1)$$

The efficiency of the compression process can also be defined as the ratio between the power requirement in isothermal compression P_{isoth} already defined and the actual power P_{actual} at the shaft of the compressor. Notice that the compressor isothermal efficiency takes losses inside the whole system into account.

It is:

$$\eta_{isoth} = P_{isoth}/P_{actual}$$

In addition, a so-called polytrophic efficiency, which can be considered as the true aerodynamic efficiency of the compressor itself with ideal gas, is introduced by developing the classic relationships. Then:

$$\eta_p = (n/(n - l))/(k/(k - 1))$$

The polytrophic efficiency of a compressor is always greater than the adiabatic efficiency already defined.

Table 11.1 Theoretical values of power requirements for different air compression processes (1 Sm³/min = 35.3 Sft³/min input air in standard atmospheric conditions, 15.6 °C, 60 °F and 0.1013 MPa, 14.5 psi; see footnotes to Table 2.5)

Outlet pressure (absolute)		Compression ratio	Isothermal 15.6 °C = 60 °F	Polytropic $n = 1.3$			Adiabatic $k = 1.4$		
				Power (kW)	Outlet temperature		Power (kW)	Outlet temperature	
MPa	psi		Power (kW)	(kW)	°C	°F	(kW)	°C	°F
0.2	29	1.97	1.15	1.24	65	148	1.27	78	172
0.3	44	2.96	1.83	2.08	98	208	2.15	121	249
0.4	58	3.95	2.32	2.73	123	254	2.84	154	310
0.5	73	4.94	2.70	3.26	144	292	3.42	182	360
0.6	87	5.92	3.00	3.71	162	324	3.91	207	404
0.7	102	6.91	3.26	4.11	178	352	4.36	228	443
0.8	116	7.90	3.49	4.47	192	378	4.76	248	478
0.9	131	8.88	3.69	4.80	205	401	5.12	266	510
1	145	9.87	3.87	5.09	217	422	5.46	282	540
1.1	160	10.86	4.03	5.37	228	442	5.77	298	568
1.2	174	11.85	4.17	5.63	238	460	6.07	312	594
1.3	189	12.83	4.31	5.87	247	477	6.34	326	618
1.4	203	13.82	4.43	6.10	256	493	6.61	338	641
1.5	218	14.81	4.55	6.31	265	508	6.86	351	663

All pressures are absolute; relationships for temperature calculations use absolute temperatures (see Sect. 11.2)
1 m³/min = 35.3 ft³/min = 0.0167 m³/s = 0.588 ft³/s
To convert from the actual volume to the standard volume see footnote to Table 2.5
Practical values of driver input power (usually electric) are 30–60 % higher than values of theoretical processes

Table 11.1 shows the theoretical values of kW needed for an inlet air flow of 1 Sm³/min (35.3 Sft³/min) in standard atmospheric conditions (60 °F, 15.6 °C; 0.1 MPa, 14.5 psi; see footnote to Table 2.5) and the values of final temperatures with different compression processes (isothermal, adiabatic, polytrophic).

If the efficiency of the drive (90 %) and other losses such as the dryer and no-load operation of the compressor are considered, typical values are 0.11–0.14 kWh per 1 Sm³ of inlet air in standard atmospheric conditions (0.0031–0.004 kWh per 1 Sft³) to compress at 0.8 MPa (116 psi). There follows that a typical power range between 6.5 and 8.5 kW is needed per 1 Sm³/min (0.19–0.24 kW per 1 Sft³/min).

11.3   Auxiliary Equipment

In practice, a compressor system includes auxiliary equipment which is essential for good operation.

Standard equipment for reciprocating compressors includes intake filter, aftercooler, and very often an air receiver.

Dynamic compressors require an inlet filter, no receiver, and sometimes an aftercooler.

Every compressor must be equipped with an air cleaner to supply adequate cool, clean and acid-free air at all times. If the air cleaner has to be placed at a distance from the compressor, care must be taken to provide a suitable suction line to the compressor.

Coolers can be classified as precooler, intercooler, and aftercooler depending on their functions. They are basically heat exchangers designed to reduce gas temperature before, during, or after compression.

Aftercoolers condense and remove moisture from compressed air before it enters the distribution system. They include condensate traps and can be air-cooled or watercooled. The latter are the better. Cooling water must be clean, cool, and soft. Cooling water generally removes 15–40 % of the total heat of compression, depending on machine size and operating characteristics. Values up to 80–90 % can be reached in intercooled centrifugal machines.

The cooling of air both reduces air temperature and volume, and condenses and removes water vapor and other condensable fluids. The shaft power requirement is reduced if cooling is effected during the compression.

After the compression, cooling may improve safety by reducing the condensable liable to cause various problems: washing away of lubricants, rusting of some parts, freezing in exposed lines during cold weather, condensation and freezing of moisture in the outlet when the expansion is considerable.

Notice that air entering the compressor at the first stage is a mixture of dry air and water vapor. The humidity ratio (mass of water vapor/mass of dry air) varies with the temperature; if the mixture temperature is reduced at constant mixture pressure, the water vapor state moves to the dew point at constant partial pressure and at constant humidity ratio. Then, if cooling goes on, condensation occurs and the final temperature is lower than the initial temperature. The system will consist of a saturated mixture of dry air and water vapor at partial pressure lower than the initial one because the mole fraction of water vapor in the final state is less than the value at the initial temperature. A fuller explanation of the basic principles of air treatment and the definition of dew-point temperature are given in Sect. 13.2.

In application at 0.7–0.8 MPa, which is a typical operating pressure in industrial plants, a dew point at 10–1.5 °C (50–35 °F) is generally adequate, depending on the humidity ratio of the outlet air.

If compressed air at this pressure never goes below the dew point after leaving the dryer, there will be no further condensation. When pressure is reduced before final use, the dew point corresponding to the reduced pressure is lower than the previous one.

Chemical drying and absorbing systems are dried-air or dehumidification systems commonly used instead of cooling equipment. Notice that all these systems are intended to remove water vapor and not lubricant oil in vapor state. If absolutely oil-free air is required, non-lubricated systems are the best.

Cooling systems use refrigerating compressors (see Chap. 12) and may use direct expansion (cold refrigerant gas expands in the air/refrigerant exchanger) or indirect expansion by using chilled water as a medium to refrigerate air. In the latter case, regenerative systems can conveniently be used: inlet air is partially cooled by discharge air, thus reducing the power of the refrigerating compressor and reheating the outlet air to the distribution line.

Advantages of reheating discharge air are the air volume is increased and thus less free air is required to move the equipment, besides, the possibility of condensation down the dew point along the distribution lines is further reduced.

Air receivers are used with reciprocating compressors. Dynamic compressors do not require receivers because they have capacity control that should equalize line output with demand.

Air receivers are very important additional equipment ensuring correct operation of the compressor. An air receiver absorbs pulsations in the discharge line from the compressor and smoothens the flow of air to the user lines. It works as a reservoir for storing compressed air to cope with sudden peak demands exceeding the capacity of the compressor. Compressors work between a lower and an upper limit of discharge pressure; the dimension of the storage must be selected with regard to technical and economic criteria, taking different factors into account, that is, load requirement, the compressor's rated air flow, and the demand pressure. Air receivers also allow moisture to precipitate and condense at the bottom, and prevent it from being carried along the air distribution pipeline.

If excess condensate is produced, a water-cooled aftercooler and separator should be added between the compressor and the receiver. This equipment dries and cools air, which means higher efficiency and safe operation for the compressor plant. When the water supply is scarce or too expensive, air-cooled aftercoolers are technically and economically justified.

11.4 Technical Data on Industrial Air Compressors and Control Systems

Table 11.2 lists the main properties of typical industrial compressors (shaft power, shaft speed, volume of compressed air and pressure). Efficient operation is always influenced by the following factors: intake air temperature (as cool as possible), minimum system pressure, absence of leaks along the pipeline, heat recovery, and selection of compressors appropriate in type and size for each application.

The intake air temperature must be kept as low as possible to reduce the work of the compressor which varies with the volume of air linearly dependent on the temperature. Table 11.3 shows how the power requirement varies with air temperature.

Table 11.2 Characteristics of typical medium-power air compressors

Type of compressor	Power (kW)	Speed (r/min)	Volume flow (Sm³/min)[a]	Absolute pressure		Compressor ratio	Isothermal power (15.6 °C) (kW)	Isothermal efficiency (%)
				MPa	psi			
Rotary air-cooled	20	2,950–3,550	2.77	0.79	115	7.8	9.6	48
	56	2,950–3,550	9.76	0.45	65	4.4	24.4	44
	112	1,450–1,775	20	0.80	116	7.9	69.7	62
	146	1,450–1,775	23	0.85	123	8.4	82.5	57
	162	1,450–1,775	23	1.00	145	9.9	88.8	55
Rotary water-cooled	20	2,950–3,550	2.77	0.79	115	7.8	9.6	48
	56	2,950–3,550	9.91	0.79	115	7.8	343	61
	90	1,450–1,775	17.8	0.79	115	7.8	61.6	68
	298	1,450–1,775	58.6	039	115	7.8	202.8	68
Reciprocating	56	400–450	9.9	0.79	115	7.8	34.3	61
	165	550–600	23	0.90	131	8.9	84.7	51
	300	550–600	64	0.79	115	7.8	221.5	74

[a]Standard atmospheric conditions 15.6 °C, 60 °F, 0.1013 MPa, 14.5 psi 1 Sm³ = 35.3 Sft³

Table 11.3 Power saving versus air-inlet temperatures at constant inlet pressure and compression ratio (consumptions are compared with power requirements in atmospheric conditions)

Temperature of air inlet		(a) Inlet volume to deliver	
°C	°F	1 Sm^3 (m^3)	Power saving (%)
0	32	0.946	5.40
5	41	0.963	3.67
10	50	0.981	1.94
15	59	0.998	0.21
15.6	60	1.000^a	0.00
20	68	1.015	−1.52
25	77	1.033	−3.26
30	86	1.050	−4.99
35	95	1.067	−6.72
40	104	1.085	−8.45
45	113	1.102	−10.18
50	122	1.119	−11.91
a	b	$c = (a + 273.15)/(15.6 + 273.15)$	$d = (1 - c) \times 100$

For every 5 °C (9 °F) rise in inlet temperature, a 1.7 % rise in energy consumption will be needed to achieve equivalent output

(a)Standard atmospheric conditions 15.6 °C, 60 °F, 0.1013 MPa, 14.5 psi; 1 Sm^3 = 35.3 Sft^3

> End users may require different pressures depending on the characteristics of the equipment; as a general rule, the lower the system pressure, the lower the compressor shaft power which varies more than linearly. Significant energy saving can thus be obtained by reducing the system pressure to the right value. When various end user machines have different power requirements, a typical approach is to install compressors which can meet the highest pressure requirement and to reduce the pressure near the low-pressure users. An alternative approach is to divide the load into different groups, on the basis of pressure needs, and to install different compressors.

Compressors are used with air- or water-cooling equipment depending on the availability of refrigerating media and on their temperatures all the year round. Air cooling is suggested for low- and medium-power sizes (less than 100 kW). Heat recovery can conveniently be applied, particularly with large water-cooled units (see Sect. 12.4).

> Positive-displacement compressors can be cooled by water which can reach a maximum temperature of 35–40 °C (95–104 °F); the use of this water for space heating is linked to the water/air exchangers which must be properly designed and equipped with forced ventilation systems. During hot seasons, heat must be discharged by cooling systems.

Screw compressors are oil-flooded compressors where oil is injected to seal clearances, absorb the heat of compression, prevent metal contact between rotors, and lubricate the bearings. Oil separation is performed immediately after compression. The oil and air mixture passes through the discharge valve to an oil reservoir, or a combination of an oil reservoir and an air receiver. The air then passes through a separator before flowing to the discharge line where most of the remaining oil is removed. Oil temperature controls are installed in all units to keep the injected oil within a range 54–71 °C (130–160 °F).

> **In screw compressors, water- or air-cooled heat exchangers are usually installed; recoverable heat temperature can be 60–80 °C (140–160 °F).**

11.5   The Control of Air Compressors

The output of compressors must be controlled to match the system demand. The parameter to control is usually the discharge pressure. A range of variation of about 10 % between discharge pressure and end user demand is generally accepted.

1. Reciprocating compressors. Automatic start-and-stop control in which compressors run at full load for a period and then are stopped can be used if starting unloaders are applied and if it is not started too often.

   Constant-speed control is performed by different methods (intake throttling, external bypassing, inlet valve loading).

   Variable-speed control, which is the ideal control for positive-displacement compressors, can conveniently be achieved by installing variable-speed drives, particularly a.c. drives for safety reasons. Variable-speed drives also provide a soft-start feature with minimum starting current, thus avoiding damage to the motor through frequent starting. This allows the compressor to stop completely and start again more frequently than in installations where electric motors are fed directly from the line;

2. Screw compressors. Automatic start-and-stop control can be used if an automatic start unloading system is installed. To unload the compressor and thus reduce the power required, convenient systems are those which include closing the inlet valve, releasing oil reservoir pressure, and reducing oil flow to the compressor. Variable-speed drives can also be used;

3. Centrifugal compressors. There are three main methods of controlling centrifugal compressors: variable speed, blowoff of unwanted output, and intake throttling.

   Speed is varied by means of variable-speed drives, either electrical or steam or gas turbine depending on the size of the system. Power decreases following speed reduction (see also the Sect. 10.3 on pumps in Chap. 10). Blowoff and recirculation control can be used to release any excess air into the atmosphere or to throttle the excess air to intake pressure and put it through the compressor again.

Intake throttling reduces discharge pressure for a given intake volume or reduces volume for a given discharge pressure. This method is widely used with constant-speed machines.

For large units, adjustable inlet guide vanes are installed ahead of the first impeller. The power saving is less than with variable-speed drives, but it is still significant;

4. Axial compressors. Adjustable vanes are typically used on the several early stages of axial compressors, since they are very sensitive to the gas angle attack and have a very short operating range.

> The variable speed drive compressor is widely used to regulate the air flow according to the end user need. It is generally installed in compressed air plant together with others compressors in order to smooth the flow diagram and reduce the no-load operating hours of fixed speed compressors. The percentage ratio between no-load and total operating hours should not exceed 10 %: in this case a value of 0.11 kWh per 1 Sm^3 (0.0031 kWh per 1 Sft^3) of inlet air in standard atmospheric conditions to compress at 0.8 MPa (116 psi) can be achieved.

11.6 Compressed-Air Systems: Distribution Lines and End Users

11.6.1 The Distribution Lines

The aim of the distribution system is to carry compressed air from the compressor to the end users at the required pressure, generally in the range 0.5–0.8 MPa (73–116 psi). The speed generally ranges from 5 to 10 m/s (11.4–32.8 ft/s); the shorter the line, the higher the speed.

Most systems consist of a main line from which secondary lines go to different production areas where individual feeder lines carry the compressed air to the end users.

For heavy intermittent demands, air receivers installed as near as possible the demand point can reduce pressure losses.

Pressure losses must be considered in selecting the pipe sizes by using tables and charts from specialized literature.

Pressure losses range between 40 Pa/m and 150 Pa/m (between 1,768 and 6,631 psi/ft) with an air-flow speed from 5 to 10 m/s (16.4–32.8 ft/s). Higher speeds would provoke higher losses. Leaks along the pipeline may cause significant losses. Notice that a hole with a diameter of 5 mm (0.02 ft) loses at least 5 kW of electric input power (with a leak of roughly 1 Sm^3/min, 35.3 Sft^3/min).

Table 11.4 lists values of losses in typical operating conditions.

A useful test is to shut down all the equipment needing compressed air and to measure the compressor's power requirement: the resulting value represents the system leaks because no useful work is delivered to the end users.

The efficiency of the distribution line between two sections A and B (see Fig. 11.3) can be calculated as follows:

$$\eta = (\ln(p_B/p_1)/\ln(p_A/p_1)) \times (100 - \text{leak}\%)$$

where p_1 = compressor inlet pressure (usually the atmospheric pressure), p_A = line upstream pressure or discharge pressure from the compressor (section A), p_B = line downstream pressure (section B, $p_B < p_A$), $p_A - p_B$ = pressure drop between sections A and B, leak% = percentage of flow leaks between sections A and B.

Typical values of the flow leaks range from 10 % to 30 % depending on the operating conditions and on the maintenance of the pipeline; the total efficiency of the distribution line ranges from 80 % to 60 %.

Figure 11.3 shows the energy balance of a complete system: electric input energy, output energy as heat (losses and recovered heat), compressed air, pipe distribution losses (leaks and pressure losses), and end users.

11.6.2 The End Users

The end users of compressed air for power purposes can be classified as positive-displacement machines, where a piston is given a reciprocating motion within a cylinder by alternate feeding of the air supply from one side of the piston to the other, and rotating machines where pistons or turbines or blades transform the energy accumulated in the compressed air into a rotating motion.

Table 11.5 lists typical values of air consumption for different tools and other machines. A coefficient with a coefficient which takes into account the discontinuous operating mode should be used. In practice, the capacity of the air compressors is equal to the sum of the single end user air consumptions with a load factor of at least 20 %.

Table 11.4 Pressure and leakage losses in air mains and related power in typical operating conditions

Nominal bore (mm)	Air flow in standard conditions (Sm³/min)	Air flow at 0.8 MPa (absolute) (m³/min)	Air speed at 0.8 MPa (absolute) (m/s)	Pressure drop per 100 m		Equivalent power		
				MPa	psi	Pressure losses (100 m) (kW)	Leakage losses (kW)	Pipe efficiency (%)
40[a]	11.9	1.5	19.9	0.07	10.15	1.7	4.1	86
50	9.2	1.2	9.9	0.015	2.18	0.3	3.2	89
65	15.8	2.0	10.1	0.01	1.45	0.3	5.5	89
80	21.1	2.7	8.8	0.006	0.87	0.3	7.3	90
100	29.0	3.7	7.8	0.005	0.73	0.3	10.1	90
125	46.1	5.8	7.9	0.004	0.58	0.4	16.1	90

Flow leaks through an orifice can be calculated by using the Bernoulli equation for the flow of compressible gas and by considering the process as adiabatic:

$$\frac{k}{k-1} \times \frac{p}{\rho} + \frac{1}{2} \times v^2 = \text{constant}$$

The air speed in the pipe is negligible in comparison with that immediately outside the orifice (typical ratio between the two speeds is roughly 30–40)

It follows:

$$m\,(\text{kg/s}) = C_d \times A \times \rho_i \times \sqrt{\left[2 \times \left(\frac{k}{k-1}\right) \times \frac{p_i}{\rho_i} \times r_p^{2/k} \times \left(1 - r_p^{(k-1)/k}\right)\right]}$$

C_d = discharge coefficient = 0.96 (typical value), A = orifice area (m²), ρ_i, p_i = air density (kg/m³) and pressure (Pa) in the pipeline, k = adiabatic coefficient = 1.4, r_p = ratio between the external and the internal pressure (p_o/p_i = atmospheric pressure/p_i)

Data for calculation (e.g., hole diameter 0.005 m)

$p_o = 0.1013$ MPa, $\rho_o = 1.23$ kg/Sm³, $p_i = 0.8$ MPa

$\rho_i = 1.23 \times \frac{0.8}{0.101}$ kg/m³ (see also footnote of Table 2.5)

$A = \frac{\pi}{4} \times \left(5 \times 10^{-3}\right)^2$ m²

$r_p = 0.1013/0.8$, $m = 0.0212$ kg/s

$Q = \left(\text{Sm}^3/\text{min}\right) = m\,(\text{kg/s}) \times \frac{60}{1.23} = 1.034$ Sm³/min   (see Table 11.1 for related power)

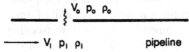

---

[a]Anomalous condition due to a high air speed

standard atmospheric condition 15.6 °C, 60 F, 0.1013 MPa, 14.5 psi

compression ratio equal to 7.9; m³ air flow at 0.8 MPa = Sm³/7.9

percentage of flow leaks assumed equal to 10 % of the pipe air flow

equivalent power is evaluated by the isothermal compression formula. Practical values are higher (see footnote in Table 11.1)

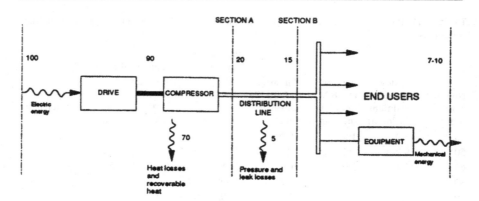

Fig. 11.3 Energy flow from input section to mechanical energy end user section in a compressed-air system

Table 11.5 Air requirements of compressed-air end user machines (absolute 0.7 MPa, 101.5 psi; 100 % load factor)

Type of machine	Air flow at 0.7 MPa = 101.5 psi		Air flow referred to standard conditions
	m³/min	m³/s	Sm³/min
Screw driver	0.15–0.3	0.0025–0.005	1–2
Concrete vibrator	1–2	0.016–0.032	7–14
Paint spray	0.2–0.5	0.003–0.0086	1.4–3.5
Steel drills	0.5–2	0.008–0.032	3.5–14
Chipping hammers	0.35–1	0.006–0.016	2–7

The power of an end user reciprocating motor depends mainly on the mean pressure differential across the air inlet and outlet, the displacement of the piston through the cylinder at maximum stroke and the number of strokes per unit of time. The degree to which air can be expanded is determined by air moisture freezing which may block the exhaust passages with ice, and not by the efficiency of the system which would impose high expansion ratios. Freezing can happen with expansion ratios as small as 2 to 1.

The work diagram of a piston motor includes two phases: a constant pressure phase (air at the distribution-line pressure enters the cylinder through valves and moves the piston, thus performing work) followed by a limited expansion (inlet valves are closed and the piston cylinder performs work by using the energy delivered during the expansion). The first phase is generally the major one; air consumption can be reduced for the same output work by cutting off the air entering as early as possible (first phase), and then allowing expansion (second phase). The opposite of the compression phase, this, is always a small part of the total work in the end user equipment.

The energy balance already introduced for the compression process can be used for end user processes.

> End user requirements are generally expressed in m^3/min if only the constant-pressure work is considered Power (W) can be calculated approximately as follows:
>
> $$air\ flow\ (m^3/s) \times end\ user\ pressure\ (Pa)$$
>
> Higher values can be obtained if the expansion work (the opposite of the compression work) is added.

11.7   Energy-Saving Investments

As discussed in Sect. 11.2, energy can be saved by improving the overall efficiency of the system.

Typical suggestions are:

- Discharge pressure must be kept as low as possible, depending on the real end user needs. Reduction of pressure by throttling valves must be avoided;
- Compressed-air temperature must be lowered in order to reduce the volume flowing through the compressor;
- Leaks along the pipeline must be carefully avoided by proper maintenance;
- The no-load operating hours must be kept as low as possible by using variable speed drives;
- Heat recovery from cooling systems, particularly for high-power systems, is always advisable;
- Proper sizing of the compressors, fixed and variable speed drive, and its related equipment allows operation at high efficiency.

11.8   Practical Examples

Example 1 Reduction of the compressor discharge pressure to the real end user needs.

A factory has an air-compressor plant which works at a discharge pressure of 0.9 MPa (130.5 psi). Most equipment may correctly operate at 0.6–0.65 MPa (87–94.3 psi), except one which needs 0.8–0.85 MPa (116–123.2 psi) with a small air flow requirement (roughly 10 % of the total air flow). Throttling valves are used to reduce the pressure to the end user requirements.

Evaluate the energy saved if the air-compressor plant operates at lower pressure, 0.7 MPa (101.5 psi), to supply the main equipment. The user at 0.8–0.85 MPa (116–123.2 psi) will be supplied with a separate air compressor.

Table 11.6 Energy saving associated with a reduction in the delivery pressure from 0.9 MPa (130.5 psi) to lower values; inlet air flow rate 30 Sm³/min (1,059 Sft³/min) in standard atmospheric conditions

			Reference plant	New plant (1) + (2)	
			Ref	(1)	(2)
a	Discharge pressure	MPa	0.9	0.7	0.9
$b = a/0.006895$		psi	130.5	101.5	130.5
c	Inlet air flow	Sm³/min	30	27	3
$d = c/60$		Sm³/s	0.5	0.45	0.05
$e = a/0.1013$	Compressor ratio		8.89	6.9	8.89
f	Electric input power[a]	kW	184	147	18.44
$g = f(\text{Ref}) - f(1)$	Power saving (1)	kW		37.54	
$g' = g \times 100/f(\text{Ref})$		%		20.35 %	
$h = f(2)$	Additional power (2)	kW			18.44
i	Working hours	h/year	5,000	5,000	5,000
$1 = (g - h) \times i$	Energy saving	kWh/year	95,492		

Theoretical isothermal power = $101.325 \times d \times \ln(a/0.1033)$
[a]Electric input power = theoretical isothermal power/0.6 (see Sect. 11.2)

Table 11.6 shows the energy saving associated with a reduction of the discharge pressure from 0.9 MPa (130.5 psi) to 0.7 MPa (101.5 psi) for the main equipment (roughly 20 % energy saving) and the additional energy required to supply the higher-pressure end user with a separate compressor.

For 5,000 h/year, the energy saved is 95,492 kWh/year.

Economic evaluation is shown in Table 19.4.

Example 2 Heat recovery from a screw compressor.

A screw compressor works between inlet air at atmospheric pressure and outlet air a discharge pressure of 0.7 MPa (101.5 psi). The air is cooled by oil injected in the rotor of the compressor. The outlet air temperature is relatively low (55–65 °C, 131–149 °F) and no additional cooling is needed.

The basic operating data are as follows:

Screw compressor
- Inlet air in standard atmospheric conditions (0.1 MPa and 15.6 °C, 14.5 psi and 60 °F)
- Air discharge pressure 0.7 MPa (101.5 psi)
- Air volume flow rate equal to 9.3 Sm³/min (328.4 Sft³/min)
- Mechanical input power 55 kW
- Outlet air average temperature 55 °C (131 °F)

Table 11.7 Energy saving associated with heat recovery from a screw compressor

Inlet air flow in standard atmospheric conditions 9.3 Sm³/min (328.4 Sft³/min)			
Discharge pressure 0.7 MPa (101.5 psi)			
a	Mechanical input power	kW	55
$b = a/0.9$	Electric input power[a]	kW	61.1
$c = 0.65 \times a$	Heat recovery[b]	kW	35.75
$d = c \times 3,600/1.055$		Btu/h	121,990.5
$e = c/0.85$	Equivalent fuel[c]	kW	42.1
$f = e \times 3,600/1.055$		Btu/h	143,518.2
$g = e \times 3,600/41,860$		kgoil/h	3.6
h	Working hours	h/year	5,000
$i = g \times h/1,000$	Energy saving	TOE/year	18.1

LHV of oil 41,860 kJ/kg
[a]Electric motor efficiency equal to 0.9
[b]Heat recovery assumed equal to 65 % of mechanical input power
[c]Boiler efficiency 0.85 (Lower Heating Value as reference)

Recovery system
- Heat exchanger oil/water
- Inlet oil temperature 95 °C (203 °F)
- Water volume flow rate 1.6 m³/h
- Water inlet temperature 60 °C (140 °F)
- Water outlet temperature 80 °C (176 °F)
- Working hours 5,000 h/year.

Table 11.7 lists basic data for the evaluation of the energy saved when water is used for space heating and other services inside the factory.

The energy saved is equal 18.1 TOE/year.

The economic evaluation is shown in Table 19.4.

Notice that the results of this example can easily be extended to other similar compressors with oil injection or to compressors cooled by means of air/water or air/air heat exchangers.

Facilities: Cooling Systems

12

12.1 Introduction

Refrigeration systems for process application and air conditioning convert work into a flow of heat from a process source or from a refrigerated space to an environmental sink. The main types can be described as vapor compression, absorption, and Brayton cycles.

Vapor compression systems are the commonest; for heat transfer they use the condensation and evaporation of a volatile medium (refrigerant or coolant) such as ammonia, light hydrocarbon, or fluorocarbon. The compressor driver supplies the input power. The basic principles of compressor operation have been discussed in Chap. 11.

Absorption systems need a high-temperature source as main power input and mechanical power only as additional power to drive a pump. Ammonia-water and lithium bromide-water systems are generally used.

Brayton cycles are used for gas refrigeration systems, where a change in phase does not occur. They are chosen to achieve very low temperatures for the liquefaction of air and other gases and for specialized applications such as aircraft-cabin cooling.

12.2 Basic Principles of Vapor-Compression Systems

Figure 12.1 shows the basic scheme of a vapor-compression refrigeration system, where power flows and heat transfers are also pointed out (they are positive in the direction of the arrows).

G. Petrecca, *Energy Conversion and Management: Principles and Applications*,
DOI 10.1007/978-3-319-06560-1_12, © Springer International Publishing Switzerland 2014

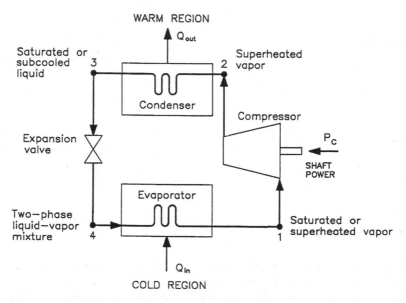

Fig. 12.1 A vapor-compression refrigeration system

The main components are the following:
- **Evaporator where the heat transferred from a process source or from a refrigerated space (cold region) results in vaporization of the refrigerant (also called coolant)**
- **Compressor where the gas refrigerant is compressed to a relatively high pressure and temperature**
- **Condenser where the refrigerant condenses and the resulting heat is transferred from the refrigerant to the cooler surroundings (warm region)**
- **Expansion valve where the liquid refrigerant expands to the evaporator pressure and leaves the valve as a two-phase liquid–vapor mixture**

As a first step, let us consider an idealized reversible cycle which represents an upper limit to the performance of the vapor-compression refrigeration cycle. This means that there are no frictional pressure losses, and that the refrigerant flows at constant pressure through the condenser and the evaporator. In addition, the compression is assumed to be an isentropic process and the expansion through the throttling valve an adiabatic process.

The main difference between the idealized and the actual cycle is in the compression process. In the latter case it is only adiabatic and the entropy increases. Pressure drops, which always occur in the cycle, are generally ignored.

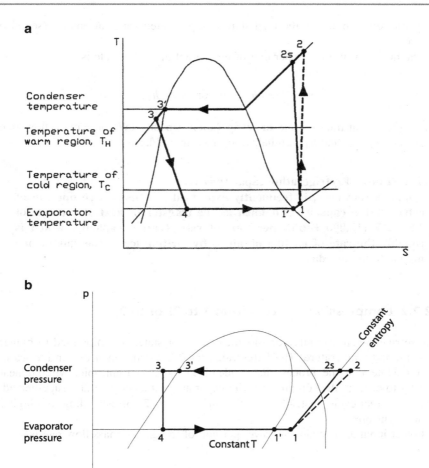

Fig. 12.2 Diagrams of vapor-compression refrigeration cycle: (**a**) Temperature-entropy diagram, (**b**) pressure-enthalpy diagram

A temperature-entropy or a pressure-enthalpy diagram can be used to represent the cycle. Both are shown in Fig. 12.2a, b.

The four phases of the cycle are described below.

12.2.1 Vaporization (Process from 4 to 1′ or 1)

When the refrigerant, as a two-phase liquid–vapor mixture, passes through the evaporator, it absorbs heat from the cold space (cold region) and changes into saturated vapor at point 1′. This happens at constant pressure and temperature. Evaporator temperature must be lower than cold space temperature (cold-region temperature). In practice, the final state is superheated vapor at point 1 (see also Sect. 6.4).

The rate of heat transfer per unit of refrigerant mass flow rate is

$$\frac{Q_{\text{in}}}{m} = h_{1'} - h_4 \text{ or } h_1 - h_4$$

where Q_{in} = heat-transfer rate or refrigeration capacity (W), m = mass flow rate (kg/s), and $h_{1'}$, h_1, and h_4 = enthalpy per mass unit (J/kg).

> **The SI unit of refrigeration capacity is W.**
> **Capacity, however, is frequently expressed in tons, where one ton of refrigeration capacity is defined as the transfer of heat at the rate of 3.52 kW (12,000 Btu/h) per tons of refrigeration capacity, which is roughly the rate of cooling obtained by melting ice at the rate of one metric tonne per day.**

12.2.2 Compression (Process from 1 to 2s or to 2)

The refrigerant, in saturated or superheated vapor state, is compressed to higher pressure and temperature. The latter value must be higher than that in the warm region. Note that the process in an idealized cycle is isentropic, without heat transfer to or from the compressor. The vapor at point 2s is generally superheated.

In an actual cycle, instead of 2s, the final point is 2 corresponding to a higher value of entropy.

Power input as mechanical power per unit of refrigerant mass flow rate is

$$\frac{P_{\text{c}}}{m} = h_{2s} - h_{1'} \text{ or } h_2 - h_1$$

where P_{c} = input power as mechanical power at the shaft of the compressor (W), and h_2, h_{2s}, $h_{1'}$, and h_1 = enthalpy per mass unit (J/kg).

The effect of irreversible compression in the actual cycle can be taken into account by introducing isentropic compressor efficiency between point 1 and point 2:

$$\eta_{\text{c}} = \frac{h_{2s} - h_1}{h_2 - h_1}$$

In addition, mechanical efficiency must be introduced (friction losses, auxiliary power such as lubricant oil pump, etc.). Notice that in hermetic compressors (in which the electric motor is placed in the same crank as the compressor), widely used in small-medium-sized equipment, the heat related to motor losses must also be introduced in the thermodynamic cycle.

Because of the abovementioned efficiencies, the shaft power is then greater than in ideal cycles by a factor of 1.3–1.4.

12.2.3 Condensation (Process from 2 to 3′ or 3)

The refrigerant passes through the condenser where it changes from superheated state to saturated vapor state, and then condenses by transferring heat to the surrounding medium (warm region) which must be at a lower temperature than the refrigerant. The final state is saturated liquid (3′) or subcooled liquid (3).

The rate of heat transfer from the refrigerant per unit of refrigerant mass flow rate is

$$\frac{Q_{out}}{m} = h_2 - h_{3'} \ \text{ or } \ h_2 - h_3$$

where Q_{out} = heat-transfer rate to the warm region (W), and h_2, h_{2s}, $h_{3'}$, and h_3 = enthalpy per mass unit (J/kg).

12.2.4 Expansion (Process from 3′ or 3 to 4)

The refrigerant, in the saturated or the subcooled liquid state, expands through the expansion valve to lower pressure and temperature in a two-phase liquid–vapor state. During this process, enthalpy remains constant because vaporization occurs by subtraction of heat from the liquid so that the temperature decreases.

Point 4 represents the end of the cycle, after which it starts again.

In the complete cycle, the net input power is the mechanical-compressor input power (or mechanical and electric in hermetic compressors), since the expansion valve does not require any power.

The coefficient of performance COP of the vapor-compression refrigeration system shown in Fig. 12.1 is

$$\text{COP} = \frac{(Q_{in}/m)}{(P_e/m)} = \frac{h_{1'} - h_4}{h_{2s} - h_{1'}} \quad \text{for idealized cycle and}$$

$$\text{COP} = \frac{(Q_{in}/m)}{(P_e/m)} = \frac{h_1 - h_4}{h_2 - h_1} \quad \text{for actual cycle.}$$

Note that the operating temperatures of the vapor-compression refrigeration cycle are mainly conditioned by the temperature T_C that must be maintained in the cold region and by the temperature T_H of the warm region where heat must be discarded. As shown in Fig. 12.2a, the evaporation temperature must be lower than the cold-region temperature T_C and the condensation temperature must be higher than the warm-region temperature T_H.

As the evaporating temperature falls further below that of the cold region and the condensation temperature rises further above that of the warm region, the coefficient of performance decreases. The warm- and cold-region temperatures represent the

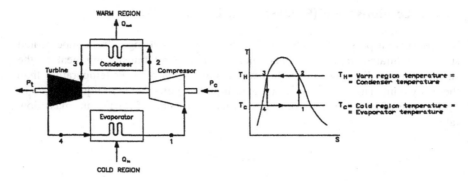

Fig. 12.3 Reverse Carnot vapor refrigeration cycle

top and bottom limits of the reverse Carnot cycle, as shown in Fig. 12.3. Of course, if the T_H or the T_C temperature varies, the coefficient of performance also varies.

> **The maximum theoretical coefficient of performance for a refrigeration cycle operating between a cold region at a temperature T_C (K) and a warm region at higher temperature T_H (K) (the reverse Carnot cycle as shown in Fig. 12.3) is**
>
> $$COP_{max} = \frac{T_C}{T_H - T_C}$$

The main differences between the reverse Carnot cycle and any other cycles may be summarized as follows:

- To achieve a significant rate of heat transfer, the difference between the refrigerant temperature in the evaporator and the cold-region temperature T_C must be equal to 5–10 °C (9–18 °F). Also, the temperature of the refrigerant in the condenser must be several degrees above the T_H temperature. Consequently, the coefficient of performance is lower than the reverse Carnot coefficient.
- The reverse Carnot cycle has a compression process with a two-phase liquid–vapor mixture or wet compression. In practice this is avoided because liquid droplets can damage the compressor, so dry compression with only saturated or superheated vapor is required. In consequence, the coefficient of performance is even further reduced below the reverse Carnot coefficient.
- The expansion process, which in a reverse Carnot cycle occurs in a turbine performing mechanical work, takes place in a throttling valve with a saving in capital and maintenance costs, but with a further reduction of the performance coefficient.

12.2.5 Practical Vapor-Compressor Cycles

In practice, vapor enters the compressor as superheated instead of saturated vapor. When superheating occurs in the evaporator or in the relative pipeline inside the space to be cooled, refrigeration capacity increases. Therefore, the mass of the refrigerant in the compressor diminishes because of the increase in the specific volume of the vapor. In addition, the temperature increases in the final compression state (point 2 instead of 2s). Attention must be paid to this value in order to maintain a good performance of the lubricant oil.

It is recommended to let the refrigerant enter the throttling valve at a temperature lower than that of saturated liquid (point 3 instead of point 3' in Fig. 12.2). In this case, the evaporation temperature downstream of the throttling valve is reached with the vaporization of less liquid.

> **The COP for a practical cycle is roughly 40–50 % of the corresponding reverse Carnot cycle COP in HVAC applications and higher when the cold-region temperature is well below the ice point (273.15 K; 0 °C; 32 °F). Practical values of COP range between 0.75 and 5.5 depending on the operating conditions, temperatures, and pressures.**
> **The lower values occur with cold-region temperatures below ice point.**

High-capacity systems with screw and centrifugal compressors usually have high COP values.

12.3 Coolant Properties and Applications

The main conditions (see typical parameter values in Table 12.1) that make a fluid a good medium for refrigerating cycles are the following:
- The evaporation enthalpy must be high and the specific heat of the fluid must be low. Consequently, during expansion less vapor will be produced to cool.
- The liquid to the evaporation temperature and refrigeration capacity will be higher.
- The temperature corresponding to the critical point must be very high and greater than the condensation temperature which depends on the warm-region temperature.
- The freezing point must be lower than the minimum evaporation temperature, which depends on the temperature of the cold region.
- The specific volume of the vapor must be as low as possible in order to minimize the size of the reciprocating compressors. On the contrary, for a proper sizing of centrifugal compressors, a significant volume flow is required to avoid diameters being too small and revolution speeds too high.

Table 12.1 Physical properties of coolants

Physical properties		R717	R22	R12	R11
Boiling point	K	240	232	243	297
	°F	−28	41.5	−21.6	74.5
	°C	−33.3	−40.8	−29.8	23.7
Freezing point	K	195.5	113	115	162
	°F	−107.9	−256	−252	−168
	°C	−77.7	−160	−158	−111
Critical temperature	K	406	369	385	471
	°F	271.5	204.8	233.6	388.4
	°C	133	96	112	198
Critical pressure	MPa	11.4	4.97	4.1	4.4
Specific volume at suction	m³/kg	0.51	0.07	0.09	0.7
Latent heat at boiling point	kJ/kg	1,367	234	234	182
Evaporator pressure	MPa	0.24	0.29	0.18	0.02
Condenser pressure	MPa	1.16	1.19	0.74	0.12
Compression ratio		4.94	4.03	4.07	6.25
Specific heat ratio c_p/c_v		1.187	1.19	1.14	1.12
Net refrigeration effect	kJ/kg	1,100	163	117	156

Average evaporation temperature 258 K, 5 °F, −15 °C
Average condensation temperature 303 K, 86 °F, 30 °C
R717 ammonia, R22 monochlorodifluoromethane, R12 dichlorodifluoromethane, R11 trichlorofluoromethane

- The operating pressures, which are linked to the operating temperatures, should avoid excessively low values in the evaporation process and excessively high values in the condensation process (typical values are 0.1–0.2 MPa, 14.5–29 psi for evaporator pressure and 0.8–2 MPa, 116–290 psi for condenser pressure). The evaporator pressure should be higher than the atmospheric one to avoid air input in the pipeline. Notice that a moderate pressure ratio should be guaranteed for the compressor to achieve better performance.
- Other properties such as chemical stability, corrosiveness, toxicity, solubility in water, compatibility with lubricant and other materials, and cost have also to be considered when choosing refrigerants.

> **As a general rule, centrifugal compressors are best suited for low evaporator pressures and refrigerants with relatively high specific volume at low pressure. Reciprocating compressors work in higher pressure ranges and with low specific volume refrigerants.**

The commonest coolants are ammonia, which is a natural compound, chlorofluorocarbons (CFCs), hydrochlorofluorocarbons (HCFCs), and hydrofluorocarbons (HFCs).

CFCs include R406, R408, R409, R502 (azeotropic mixture of R22 and R115), R11, R12, and R115.

HCFCs include R22 and other hydrochlorofluorocarbon gases.

HFCs include R134a, R144, and other hydrofluorocarbon compounds.

CFCs and HCFCs are harmful to the ozone layer in the upper atmosphere (they are called ozone-depleting gases) and therefore their use in the refrigeration industry has been banned in some countries.

HFCs, which are widely used as alternative coolants, are greenhouse gases (GHGs) (see also Sect. 6.11).

The main properties of coolants are reported in Table 12.1. Typical fields of application are R12 (domestic refrigeration, small plants with $T_c > -20$ °C, automotive air conditioning), R22 and R134a (food conservation, small- and medium-size space air conditioning), R502 (low-temperature food conservation), ammonia (low-temperature food conservation and freezing), and R11 (large air conditioning plants).

12.4   The Main Components of Vapor-Compression Systems

As shown in Fig. 12.1, the main components of a vapor-compression system are evaporator, compressor, condenser, and expansion valve.

12.4.1 Evaporator

The evaporator is basically an exchanger where the refrigerant, as a liquid or a liquid–vapor mixture at low temperature and pressure, evaporates by subtraction of heat from the process or from the space to be cooled.

A liquid–vapor mixture exists when the evaporator is fed immediately downstream of the expansion valve.

Two basic types of evaporator are used in the refrigeration industry: (1) the direct expansion type, in which the fluid evaporates inside tubes, suitable for direct air cooling, and (2) the flooded type (pool boiling) where the fluid boils in contact with tubes which contain the warmer liquid (as for the production of chilled water).

Attention must be paid to the formation of ice around evaporators working in air at a temperature below 0 °C (32 °F). The defrosting can be done by recirculating hot gas or by inverting the cycle or by electric resistors. The energy consumption in this phase must be taken into account when calculating the COP of the system.

12.4.2 Compressor

The compressor has the task of increasing the pressure of the refrigerant from the evaporating value to the condensing value, to which corresponds the condensation temperature. This temperature must be compatible with the surrounding or warm-region temperature, as already explained.

Compressors for refrigeration can be classified mainly as reciprocating compressors, screw compressors, and centrifugal compressors. The choice among them depends on the power range, the working temperatures, the range of demand from end users, etc. Detailed information on compressors can be found in specialized textbooks.

As a general rule, reciprocating compressors are used up to a maximum refrigerating capacity of 500–600 kW; centrifugal compressors have practically no upper limits and their application starts from a lower limit of a few hundred kW; screw compressors usually cover a range between 200 and 4,000 kW. Reciprocating and screw compressors can easily be adapted to different working temperatures and pressures. The centrifugal compressor adapts poorly to variations in pressure.

Centrifugal and screw compressors can regulate the useful power continuously; attention must be paid to the low operating limit, which always exists in centrifugal compressors, established by the surge region where instability may occur. If this limit is reached, either additional load must be placed on the machine by recycling vapor from discharge to suction or one compressor must be taken out of service. Screw compressors can regulate capacity in a wide range, between 10 and 100 %, by regulating the quantity of gas at suction.

Reciprocating compressors can regulate the useful power step by step by operating the suction valves sequentially.

Control of speed by a variable-speed drive is always possible.

12.4.3 Condenser

The condenser is basically a heat exchanger. Its main function is to condense the refrigerant from a vapor state to a liquid state (saturated or subcooled) by means of a surrounding medium (warm region) at lower temperature. With reference to Fig. 12.1 the rate at which heat (Q_{out}) is subtracted from the refrigerant is

$$Q_{out} = Q_{in} + P_c \quad (\text{W})$$

where Q_{in} = refrigerating capacity (W), and P_c = mechanical power at the compressor shaft or electric power in hermetic compressors (W).

Three main modes of operation occur in the condenser depending on the different states through which the refrigerant passes: from superheated to saturated vapor, from saturated vapor to saturated liquid, and from saturated liquid to subcooled liquid. The sum of the three terms mentioned above is Q_{out}, already introduced.

1. The heat to be discarded in the first step, from superheated to saturated vapor, is

$$c_p \times m \times (t_2 - t_k)$$

where c_p = specific heat at constant pressure of the superheated vapor, t_2 = temperature of vapor entering the condenser, t_k = condensation temperature corresponding to the working pressure, and m = mass flow rate of the refrigerant.
2. The heat to be discarded during the condensation phase is

$$m \times h_1$$

where h_1 = latent heat at working conditions (temperature and pressure).
3. The heat to be discarded during the subcooling phase is

$$c \times m \times (t_k - t_3)$$

where c = specific heat of the liquid, and t_3 = outlet temperature of the subcooled liquid.

Depending on local conditions and on technical considerations, water or air can be used as cooling media.

If water is used, condensers are refrigerant fluid/water exchangers. Water can be discarded, but it is better to cool and recirculate it by appropriate cooling equipment such as cooling towers. These systems are based on the evaporation of a small quantity of the circulating water by subtraction of heat from the remaining water, which is then cooled. Note that the temperature of the cooled water is always higher than the wet-bulb temperature that indicates the lowest value to which the circulating water can be cooled. In practice, the design temperature at the evaporating tower outlet is assumed to be at least 5 °C (9 °F) higher than wet-bulb temperature (see also Sect. 13.2).

> **When discarded water is used for cooling, in the case of outlet-inlet temperature changes of 15–20 °C (27–36 °F), practical values of water consumption range between 0.1 and 0.2 m³/kWh of input energy, generally electric energy consumed by the compressor prime mover. Higher values occur if the water outlet-inlet temperature change is smaller. When evaporating towers are used for cooling, water consumption values range between 0.01 and 0.02 m³/kWh. Electric energy consumption of auxiliary equipment (pumps, etc.) ranges between 0.05 and 0.1 kWh/kWh of input energy consumed by the compressor prime mover.**

Evaporating condensers are also used, in which condensation is due to the evaporation of external water. The condensation temperature is generally 10–12 °C (18–22 °F) higher than external air temperature.

If air is used, forced circulation of air must be employed to facilitate heat transfer in refrigerant fluid/air exchangers (air condenser). This system, which requires electric power for fans, has the disadvantage that working temperatures vary with air temperature. If the temperatures are too high, excessively high condensation temperatures may be required. As a result, the COP is reduced.

Practical values of energy consumption in air-condenser fans are 0.05–0.1 kWh/kWh of input energy, generally electric energy consumed by the compressor prime mover.

12.4.4 Expansion Valve

The expansion valve is a typical reduction valve that lowers the pressure to a value suitable for the downstream evaporator's operation. The pressure is controlled by the liquid fluid level in flooded evaporators and by the superheated vapor temperature in direct expansion evaporators (thus ensuring roughly 5–8 °C; 9–14.4 °F of superheating temperature increase at the suction line).

12.5 Cascade and Multistage Systems

In order to obtain a good coefficient of performance with very low evaporating pressure and relatively high condenser pressure, cascade and multistage cycles are generally used.

A cascade cycle, in which two vapor-compression cycles are located in series, is illustrated in Fig. 12.4. A counterflow heat exchanger is situated between the two cycles: the heat rejected during the condensation of the low-temperature cycle (cycle A) is used to evaporate the refrigerant in the second cycle which works in a higher temperature range. If necessary, two refrigerants with different saturation pressure temperature relationships can be selected to ensure the best temperature and pressure for the two temperature ranges: low evaporation temperature without excessively low pressure for the first cycle, and relatively high condensation temperature without excessively high condensation pressure for the second cycle.

Multistage compression with intercooling between the stages, achieved by means of the refrigerant itself, can also be used if high temperature differences between cold and warm regions are required.

12.6 Absorption Refrigerating Systems

Absorption refrigeration cycles requiring heat instead of mechanical energy as the main input power can be used in particular applications and in trigeneration plants (see Chap. 9). The basic scheme is shown in Fig. 12.5. Notice that this system must always operate at temperature above 0 °C (32 °F).

Fig. 12.4 Example of a cascade vapor-compression refrigeration cycle

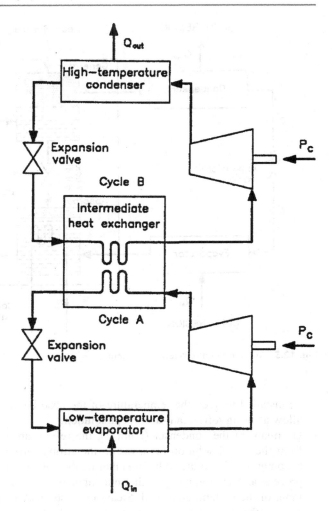

They differ from the vapor-compression cycles in two important respects:

- Downstream of the evaporator (point 1), the refrigerant in the vapor state is absorbed by a secondary substance, called an absorbent, to form a liquid solution. This solution is then pumped to a higher pressure (condenser pressure) by means of a pump which requires much less work than vapor compression does, because of the low specific volume of the solution. The mechanical input power is consequently very low.

Note that the formation of the liquid solution is an exothermic reaction. Since the amount of refrigerant that can be dissolved into the absorbent liquid increases as the solution temperature decreases, a cooling system may

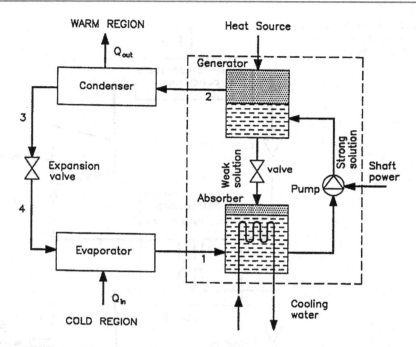

Fig. 12.5 An absorption refrigeration system

be installed to keep the temperature of the absorber as low as possible and to allow a strong refrigerant-absorbent solution.

• Upstream of the condenser (point 2), the refrigerant vapor must be retrieved from the liquid solution (as a result, the strong refrigerant-absorbent solution becomes a weak solution) before entering the condenser. This is an endothermic process and consequently a thermal input power, which represents the main input of the system, is needed. Steam or waste heat are generally used; natural gas or other fuels can be burned to provide the heat source. The choice of heat source is determined mainly by cost.

The refrigerant vapor passes to the condenser and the remaining solution, which is a refrigerant weak solution, flows back into the absorber tank through a valve without entering the condenser.

The commonest cycles use lithium bromide-water solution in which water is the refrigerant and lithium bromide the absorbent; pressures are always lower than the atmospheric value. Another type of absorption system uses ammonia-water solution in which ammonia is the refrigerant and water the absorbent.

Evaporation temperature must be always higher than 3–5 °C (37–41 °F). As a general indication, 2–1 kW of heat at temperatures between 80 and 120 °C (176–250 °F) or higher are required to produce 1 kW of refrigeration capacity. The coefficient of performance varies between 0.5 and 1, which is much lower than the corresponding coefficient for vapor-compression cycles. Lower values correspond to low-temperature hot water as heat source and higher ones to the direct use of fuel. The heat power to be discarded by means of condensing equipment (cooling tower, air condenser, etc.) is roughly 2.5 kW for a refrigeration capacity of 1 kW ($Q_{out} = Q_{in}$ + heat source).

Higher COP up to 1.2 can be obtained in cycles using a direct fired system fed by combustibles, with a two-stage absorption system.

Additional equipment is needed to improve the coefficient of performance, for example heat exchangers that allow the strong solution entering the generator to be preheated by the weak solution leaving it, and rectifiers to remove any traces of water from the refrigerant before it enters the condenser.

In most cases absorption systems use recovering heat in trigeneration plants (see also Chap. 9).

12.7 Brayton Refrigeration Cycle

The Brayton refrigeration cycle is one of the gas refrigerating systems in which the process involves no change in phase. The basic scheme of this cycle is represented in Fig. 12.6. A modified scheme incorporating an intermediate heat exchanger between the fluids leaving the two basic exchangers is the form most often used.

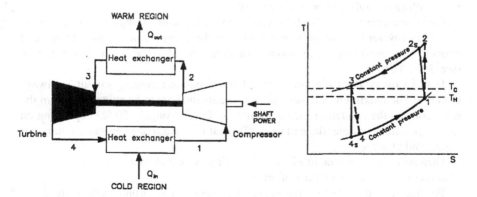

Fig. 12.6 Basic scheme of a Brayton refrigeration cycle

12.8   Energy Saving and Computer Control

The coefficient of performance is the index commonly used to compare refrigeration systems.

As explained in Sect. 12.2, the upper limit is related to the Carnot cycle in which the COP index depends only on the cold- and warm-region temperatures (T_C, T_H).

The need for a difference between the evaporation and condensation temperatures and those of the cold and warm regions (which are the T_c and T_H limiting temperatures of the Carnot cycle) together with the losses inside the components of the systems makes the real cycle less efficient than the ideal one.

In order to obtain operating conditions with the best COP, and so to ensure significant energy saving, the following suggestions can be made for both plant management and capital investment:

Plant management
• The evaporation temperature must be the highest permitted by the needs of the users. This can be achieved by setting the temperature control of the fluid to be cooled or of the suction pressure of the valve.
• The condensation temperature must be the lowest permitted by the cooling system and by the warm-region temperature.
• The load rate of the single compressors must be kept as high as possible by taking out the single machines in sequence in order to avoid low efficiency occurring at low rate.

Capital investment
• Installing a floating point for the control of the evaporating temperature according to external conditions and user needs
• Improving the efficiency of the cooling equipment to allow lower condensation temperatures
• Installing a variable-speed drive to regulate the refrigerant flow without dissipation
• Installing centralized computer control

Costs and energy can also be saved by installing cold storage systems, such as ice or cold-water or eutectic storage, to shift the energy consumption from peak hours to low-rate hours and to smooth load demand and so the system power profile (see Sect. 4.9).

Heat recovery from the hot water of the condenser cooling circuit is always possible. The maximum recoverable heat equals the heat to be subtracted from the refrigerant in the condenser; actual values are lower (roughly 50 %) depending on the temperatures of potential end users. Typical uses are space heating, sanitary hot water, and cleaning.

Heat recovery from the lubricant oil-cooling system allows higher temperatures than recovery from the condensation phase.

For high-power systems, drivers such as steam or gas turbines can be used in a cogeneration cycle.

12.9 Heat Pump Operation

A heat pump is a system operating cyclically to transform low-temperature energy into higher temperature energy by the application of external work. Both vapor-compression and absorption systems can be used as heat pumps with small modifications.

The coefficient of performance has the same meaning as in the refrigeration systems shown in Figs. 12.1 and 12.2, but input and output energies play different roles. The useful output or desired effect is the heat directed to the mass to be heated (Q_{out}) so that

$$COP = \frac{(Q_{out}/m)}{(P_c/m)} = \frac{h_{2s}-h_{3'}}{h_{2s}-h_{1'}}$$ for idealized cycle and

$$COP = \frac{(Q_{out}/m)}{(P_c/m)} = \frac{h_2-h_3}{h_2-h_1}$$ for the actual cycle.

Heat pumps can be used for space heating or for industrial processes to raise the temperature of available heat, generally waste heat at low temperature, or of ambient air or water sink, to a level acceptable to the end users. The prime mover is either an electric motor or a fuel engine.

> As a general indication, only if the COP is greater than 3 can heat pumps be regarded as an energy source economically more attractive than boilers. This depends on the overall efficiency of the prime mover, which may be a thermal or an electric motor, and on the end user's energy costs.

Primary energy saving, more than cost saving, depends on the efficiency of all the energy transformations. In the case of a fuel engine, the efficiency is the ratio between the mechanical power at the shaft and the input power as fuel in the engine itself. In the case of an electric motor, the efficiency is the ratio between the mechanical power at the shaft and the thermal input power needed in the utility power plant to produce electric energy.

As in refrigerating systems, in heat pump operation the maximum theoretical COP obtainable is that of the reverse Carnot cycle:

$$COP_{max} = \frac{T_H}{T_H - T_C}$$

One of the main disadvantages of heat pumps is the operating temperature range: too low evaporation or cold-region (heat-source) temperatures yield a low COP, if relatively high temperatures are required as useful output. Contrariwise, in industrial applications, an additional constraint is that high-temperature waste above 95 °C (200 °F) cannot be supplied directly to the heat pump because of the limits imposed by refrigerant characteristics.

Figure 12.7 shows mean heat-pump theoretical COP values versus load (warm-region) temperatures for a number of source temperatures.

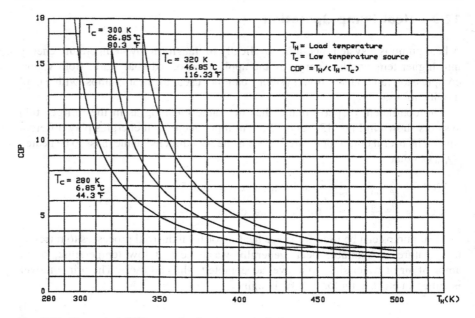

Fig. 12.7 Theoretical COP versus load temperatures for heat pumps

In HVAC applications, heat pumps have evaporators in direct contact with the outside air, which serves as energy source during the cold season. In the hot season, the same air source can provide cooling through a reversing valve. Figure 12.8 shows how the refrigerant flows inside the same components in heating and in cooling modes. Note that in the heating mode the outside heat exchanger works as evaporator (connected with the compressor inlet) and the inside exchanger works as condenser. In the cooling mode, the outside heat exchanger becomes the condenser and the inside exchanger becomes the evaporator (connected with the compressor inlet).

12.10   Practical Examples

Two examples of energy saving are here presented.

The first refers to the replacement of an absorption system by a rotating compressor system in a site where waste heat streams are not available, so that all the heat needed by the absorption system must be supplied by means of fuels.

The second concerns heat recovery from condensation and lubricant oil cooling, which is a possibility always present, but exploitable only when end users really need heat at low temperature.

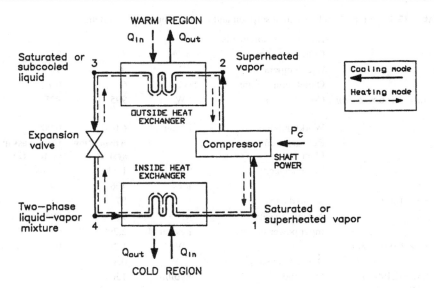

Fig. 12.8 Example of air-to-air reversing heat pump

Example 1 Replacement of an absorption system by a rotating compressor system.

An absorption system with a rated capacity of 1,744 kW (496 tons) is installed in a food-processing factory and it is fed by steam at 0.7 MPa. The actual refrigeration capacity is 995 kW (283 tons) with a steam flow rate equal to 2,299 kg/h. The replacement of this system by a rotating compressor is here examined from the energy-saving point of view, on the assumption that end-user needs are satisfied in both cases. The operating characteristics are detailed in Table 12.2 together with the technical comparison.

Economic evaluation is in Table 19.4.

Example 2 Heat recovery from condensation phase and lubricant oil cooling.

A refrigerating plant with screw compressors has a nominal capacity of 1,400 kW (397.7 tons). The input shaft power is 500 kW; the refrigerating plant works 5,000 h/year at 70 % of the rated capacity for food conservation at low temperature.

Heat recovery can be obtained in two ways: (1) from the condensation circuit by means of gas/water exchangers and (2) from lubricant oil-cooling equipment.

In case (1), hot water at 40–45 °C (104–113 °F) can be obtained if the condensation temperature is kept at 45–50 °C (113–122 °F), which is quite a high value but still acceptable for screw compressors. Hot water can be used for heating purposes in the HVAC plant of the site and the boiler-fed hot water circuit for integration, when needed.

The recoverable heat is estimated equal to 60 % of the sum of the shaft power and the refrigeration capacity at the average load, that is, 798 kW.

Table 12.2 Comparison between absorption and rotating compressor systems

			Absorption system (1)	Compressor system (2)
	End-user requirements			
a	Cold-water flow	m³/h	372	372
b	Inlet temperature	°C	9	9
c	Outlet temperature	°C	6.7	6.7
$d = a \times 4.186 \times (b-c)/3.6$	Cooling load	kW	995	995
e	Working hours	h/year	6,000	6,000
	Refrigeration Plant		**Absorption system (1)**	**Compressor system (2)**
f	Rated capacity	kW	1,744	1,744
$f' = f/3.52$		tons	496	496
$g = d \times 100/f$	% rated capacity	%	57	57
$h = d$	Actual capacity	kW	995	995
i	Input power (steam)	kgsteam/h	2,299	0
$i' = i \times 2{,}190/3{,}600$		kW	1,399	0
l	Boiler efficiency		0.8	
$m = (i \times 2{,}190/1)/41{,}860$	Input fuel	kgoil/h	150	
n	Input power (electricity)	kW	0	245
$o = h/i'$	COP		0.711	
$o = h/n$	COP			4.067
	Cooling towers and auxiliaries			
$p = d$	End user load	kW	995	995
$q = i'$	Input power (steam)	kW	1,399	
$q = n$	Input power (electricity)	kW		245
$r = p + q$	Total to dissipate	kW	2,394	1,240
s	Input electric power	kW	113	58
t	Water consumption	m³/h	7.5	3.9
	Energy consumption			
$A = m \times e/1{,}000$	Thermal energy	TOE/year	902	0
$B = (s + n) \times e/1{,}000$	Electric energy	MWh/year	678	1,818
$C = t \times e$	Water	m³/year	45,000	23,400
	Energy saving			
$A(1) - A(2)$	Thermal energy	TOE/year		902
$B(1) - B(2)$	Electric energy	MWh/year		−1,140
$C(1) - C(2)$	Water	m³/year		21,600

Notes: Useful enthalpy = steam enthalpy − condensate enthalpy = 2,190 kJ/kg steam
LHV of oil 41,860 kJ/k
All input power figures of refrigeration plant and auxiliaries are average values referred to 6,000 h/year

Table 12.3 Energy saving associated with heat recovery from a refrigerating plant

Case (1)	Recovery from condensation phase		
a	Rated refrigeration capacity	kW	1,400
b	Rated shaft power	kW	500
c	Average load factor	%	70
$d = c \times (a + b)/100$	Condensation heat at average load	kW	1,330
$e = 0.6 \times d$	Recovered heat	kW	798
$f = e/0.85$	Equivalent input power as fuel[a]	kW	938.8
$g = f \times 3{,}600/1.055$		Btu/h	3,203,568
$h = f \times 3{,}600/41{,}860$		kgoil/h	80.7
i	Working hours of recovered-heat users[b]	h/year	2,400
$l = h \times i/1{,}000$	Energy saving	TOE/year	193.8
Case (2)	Recovery from lubricant oil cooling		
a	Rated refrigeration capacity	kW	1,400
b	Rated shaft power	kW	500
c	Average load factor	%	70
$d = c \times (a + b)/100$	Condensation heat at average load	kW	1,330
$e = 0.1 \times d$	Recovered heat from lubricant	kW	133
$f = e/0.85$	Equivalent input power as fuel[a]	kW	156.5
$g = f \times 3{,}600/1.055$		Btu/h	533,928
$h = f \times 3{,}600/41{,}860$		kgoil/h	13.5
i	Working hours of recovered-heat users[c]	h/year	6,000
$l = h \times i/1{,}000$	Energy saving	TOE/year	80.7

Notes: LHV of oil 41,860 kJ/kg
[a]Boiler efficiency 0.85 (Lower Heating Value as reference)
[b]Working hours for heating
[c]Working hours for heating and other uses

In case (2), hot water at 45–50 °C (113–122 °F) can be obtained by means of a plate exchanger regardless of the condensation temperature; the discharge temperature of the lubricant oil ranges between 50 and 55 °C (122 and 131 °F), so this can be injected into the compressor at 45–50 °C (113–122 °F) with a temperature drop of 5 °C (9 °F). The recoverable heat is estimated to be equal to 10 % of the condensation heat, that is, 133 kW, much less than in case (1). Hot water can be used as in case (1) or for other users.

Energy saving in both cases has been calculated in Table 12.3 by introducing the efficiency of an equivalent boiler.

Economic evaluation is shown in Table 19.4.

Facilities-HVAC Systems

13

13.1 Introduction

Heating, ventilating, and air conditioning systems may be among the largest users of energy, depending on the activities performed in the buildings where they are installed. In commercial buildings, hospitals, data centers, and some industries (pharmaceutical, chemical, food processing, electronics) the main energy users are HVAC and lighting, while in the majority of industries most consumption is due to process users.

In typical HVAC plants, the main electric energy users are the refrigeration plant, the air distribution fans in air-handling units (AHU), and auxiliaries such as cooling towers and circulating pumps. Generally, the only user of fuel is the boiler plant (hot water or steam) for space-heating purposes in cold seasons and for re-heating after dehumidification in hot seasons. The relative importance of these energy consumptions depends mainly on the local climate and for how long heating or cooling modes are needed.

> HVAC plants are designed to provide a range of comfortable conditions most of the time for most people working in the building. It is not easy to formulate a universally applicable definition of comfortable conditions; an ambient temperature of between 23 and 25 °C (73–77 °F), a relative humidity of 40–60 %, a specific humidity of 0.008–0.012 kg of water per kg of dry air, and maximum air velocity of less than 0.15–0.20 m/s (9–12 in/min, 30–40 ft/min) are generally assumed as optimal conditions. These values are different for industry, where temperatures can be lower and air velocity higher, up to 0.5 m/s.

The need for heating, cooling, and ventilation depends on many factors such as external temperature and humidity, the number of people working in the room and their activities and clothing, the power installed (machinery and lighting), the type of

G. Petrecca, *Energy Conversion and Management: Principles and Applications*,
DOI 10.1007/978-3-319-06560-1_13, © Springer International Publishing Switzerland 2014

processes performed (e.g., textile, food processing, chemical, pharmaceutical), and the functions of the buildings (commercial center, hospital, bank, industry, etc.).

Heating and cooling requirements usually range between 50 and 100 W/m^2 and more for particular applications; considering ceiling height, typical values are 20–30 W/m^3 for residential buildings and less than 10 W/m^3 for industries without particular requirements. Of course, climatic conditions, type of activity performed, and indoor air requirements strongly influence these values.

In HVAC systems, hot water for heating is generally supplied at 4,060 °C (104–140 °F) and chilled water at 5–10 °C (41–50 °F); hot air at 30 °C (86 °F) and cold air at 13–15 °C (55–59 °F). The difference between inlet and outlet water temperatures is 10–20 °C (18–36 °F) for heating and 5–10 °C (9–18 °F) for cooling.

13.2 Basic Principles of HVAC Systems

The basic task of each HVAC system is to control the temperature and humidity of the air inside buildings and thus to ensure comfort; economics requires reaching this result with the minimum energy consumption.

In small plants, air temperature control is the simplest technique, but humidity control plays a fundamental role in ensuring comfort for occupants and the right environment for many industrial processes that require control of the vapor content in the air.

Design and operation of HVAC are based on the knowledge of the behavior of mixtures of dry air and water vapor, commonly called moist air, which can be treated as a mixture of ideal gases.

Typical states of water vapor in moist air are shown in the T,v diagram of Fig. 13.1: the dry air-water vapor mixture has a pressure p and a temperature T^* and the components have a common temperature and pressures equal to the mixture pressure p multiplied by the single component mole fraction (p_v for vapor, p_a for air; $p_a > p_v$; $p = p_a + p_v$). At typical mixture temperatures and pressures, the vapor is superheated (p_v; typical situation as at point 1).

The mixture is said to be saturated (saturated air) when the partial pressure of vapor p_v has the same value as the saturation pressure of water p_w at the mixture temperature (p_w as at point 2; typically, $p_w > p_v$).

The quantity of water vapor in moist air varies from zero in dry air to a maximum value, which depends on the pressure and temperature at which the mixture is saturated. Two different indexes of the quantity of water vapor in moist air are used: humidity ratio or specific humidity (ω) and relative humidity (Φ).

Humidity ratio or specific humidity (ω) is the ratio between the mass of water vapor (M_v) and the mass of dry air (M_a). This ratio reaches its maximum value in

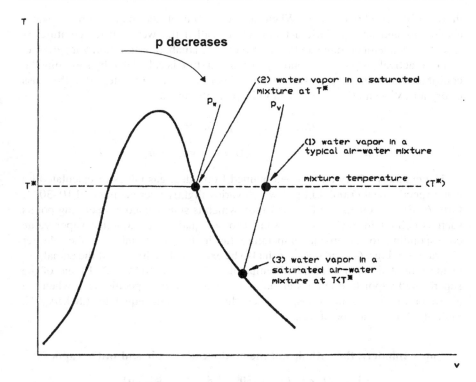

Fig. 13.1 Typical water vapor states in an air-water mixture

saturated air (practical values range from zero for dry air to a maximum value of less than 0.03 kg water per kg dry air). By using the ideal gas laws and the molecular weights of water and dry air, the humidity ratio can be expressed as the ratio between the molecular vapor weight multiplied by its partial pressure p_v and the molecular dry air weight multiplied by its partial pressure $(p - p_v)$, that is, $0.622 \times p_v/(p - p_v)$.

Relative humidity (Φ) is the ratio between the mole fraction of water vapor in a given moist air sample and the mole fraction of water vapor in a saturated moist air sample at the same mixture temperature and pressure. This ratio equals 1 (or 100 if expressed as a percentage) for a saturated mixture, so its values indicate how moist air should be treated.

Notice that air-water mixture in normal conditions has the adiabatic-saturation temperature close to the wet-bulb temperature which can be measured by a psychrometer.

The thermodynamic wet-bulb temperature is the temperature at which water, by evaporating into moist air at a given dry-bulb temperature and humidity ratio, can bring air to an adiabatic saturation while the mixture pressure is constant. The psychrometer consists of two thermometers with one bulb covered by a wick thoroughly wetted with water. When the wet bulb is placed in an airstream, water evaporates and an equilibrium temperature, called the wet-bulb temperature, is reached. This temperature can be assumed as the adiabatic-saturation temperature.

For practical purposes, the enthalpy of the mixture is calculated by summing the contributions of dry air and water vapor. This value is usually referred to the mass of dry air (Ma) and the humidity ratio is then introduced:

$$H_{mix} = M_a \times h_a + M_v \times h_v$$

$$(H_{mix}/M_a) = h_{mix} = h_a + (M_v/M_a) \times h_v = h_a + \omega \times h_v$$

The enthalpy of dry air can be obtained from ideal gas tables or calculated by using specific heat (1.005 kJ/kg K which varies slightly over the interval 10–30 °C (50–86 °F). The enthalpy of water vapor, which is superheated at operating points such as point 1 in Fig. 13.1, is assumed to be equal to the saturated vapor value corresponding to the given temperature (as in the steam table or the Mollier diagram; see Figs. 6.1 and 6.2). It can also be expressed as the sum of the enthalpies of the saturated vapor at standard temperature 273.15 K (0 °C, 32 °F) and of the superheated vapor (the latter's enthalpy is equal to vapor-specific heat, when the air-water mixture is in atmospheric conditions assumed equal to 1.8 kJ/kg K, multiplied by the temperature).

Approximately, the enthalpy of the mixture of dry air and water vapor is

$$h_{mix} = 1 \times t + \omega \times (2500 + 1.8 \times t) \quad (kJ/kg)$$

where
1 kJ/kg K is the specific heat of the dry air.
** t is the difference between the mixture temperature in K and the standard temperature (273.15 K), that is, the temperature in °C.**
** 2,500 kJ/kg is the enthalpy of the saturated steam and 1.8 kJ/kg K is the specific heat of the superheated steam when the air-water mixture is in atmospheric conditions ($p_v < p_a$; see Table 2.5).**

The humidity ratio of a moist air mixture varies with the temperature; if the mixture temperature T* is reduced at constant mixture pressure, the water vapor state moves from point 1 to the dew-point temperature (point 3 in Fig. 13.1) at constant partial pressure p_v and at constant humidity ratio. Then, if cooling continues condensation occurs and the final temperature is less than T*. The system will then consist of a saturated mixture of dry air and water vapor at a pressure less than p_v, because the mole fraction of water vapor at the final state is less than the value at the initial temperature.

The dew-point temperature of a given air sample is the temperature of saturated moist air at the same pressure and with the same humidity ratio as the air sample.

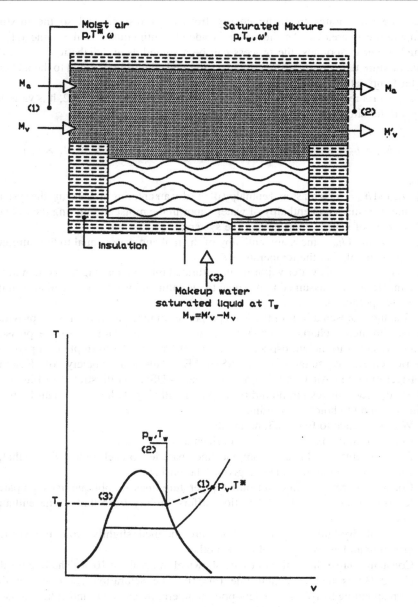

Fig. 13.2 Adiabatic saturator: Basic equipment and T–v diagram representation

By adding water in an adiabatic process to a moist airstream, the humidity ratio can be incremented as shown by the T,v diagram of Fig. 13.2. When a given mixture at known pressure and temperature (pressure p and temperature T^*) comes into contact with water at temperature T_w (the example has T^* greater than T_w), some of the water evaporates by using energy from the moist air and so the mixture temperature decreases. As the water is added continuously at the same rate at which water evaporates, the process can be assumed as an adiabatic one and the final mixture temperature, generally in a saturated condition, is equal to the makeup water temperature T_w.

A general relationship based on an energy balance is (see Fig. 13.2, it is valid for any similar situation)

$$(M_a \times h_a + M_v \times h_v) + \left[\left(M'_v - M_v \right) \times h_w \right] = \left(M_a \times h'_a + M'_v \times h'_v \right)$$

where

M_a, h_a, and $h'_a = h_a =$ mass and enthalpy of the dry air entering and leaving the system.

M_v and $h_v =$ mass and enthalpy of water vapor in the moist air entering the system.

$M'_v =$ mass of water vapor leaving the system.

$(M'_v - M_v)$ and $h_w =$ mass and enthalpy of the makeup water equal to the saturated liquid enthalpy at the temperature T_w.

$h'_v =$ enthalpy of the water vapor in the saturated mixture leaving the system at the temperature T_w (assumed to be equal to the saturated water vapor enthalpy at the T_w temperature).

Graphic representations of the main properties of moist air mixtures are provided by psychrometric charts, where sets of curves are given for a mixture pressure generally equal to the atmospheric pressure (0.1 MPa). An example of a psychrometric chart, as suggested by ASHRAE (American Society of Heating, Refrigerating and Air Conditioning Engineers—USA), is illustrated in Fig. 13.3, where the coordinates are humidity ratio and enthalpy. Other charts can be used, with a different choice of coordinates.

With reference to Fig. 13.3, notice that:
- Constant humidity ratio lines are horizontal.
- Constant enthalpy lines are oblique lines parallel to each other. The enthalpy values correspond to the above given relationships.
- Constant thermodynamic wet-bulb temperature lines are oblique lines not parallel to each other and their direction differs slightly from that of the enthalpy lines.
- Constant dry-bulb temperature lines are inclined slightly from the vertical direction and are not parallel to each other.
- Constant relative humidity lines are shown at intervals of 10 %. The line for dry air (0 %) is the horizontal line ($\omega = 0$). The saturation curve is the line corresponding to 100 %; the dew-point temperature can be found at the intersection of constant ω line with the saturation curve.
- Constant specific volume lines are straight, but not parallel to each other.

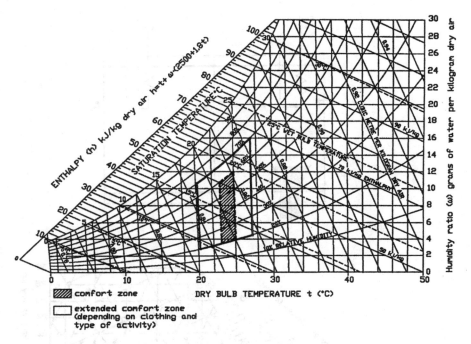

Fig. 13.3 An example of a psychrometric chart as suggested by ASHRAE

13.2.1 Typical Air Conditioning Processes

Most of the typical air conditioning processes (dehumidification, humidification, adiabatic mixing, etc.) can be described using the basic principles discussed above and psychrometric charts.

> When necessary to lower the relative humidity, dehumidification is generally implemented by a cooling section followed by a re-heating section. There are also other methods of dehumidification, such as chemical absorption by salts.

In a dehumidification process by a cooling section (see Fig. 13.4), the moist air stream (point 1) is cooled below its dew-point temperature, so that the initial water vapor condenses. A saturated moist air mixture (with relative humidity 100 % in ideal operating conditions and lower humidity ratio) leaves the dehumidification section (point 2) at the saturation temperature below the dew point (the final partial pressure of water vapor at point 2 is lower than the initial partial pressure of water vapor in the mixture, because condensate has been removed). This moist airstream must generally be re-heated in a following section in order to bring it to a condition most occupants would consider comfortable. This last phase occurs at constant

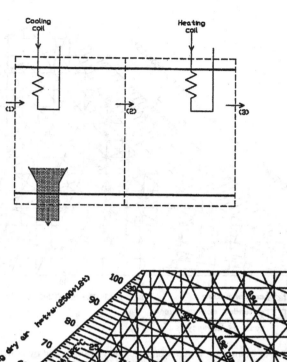

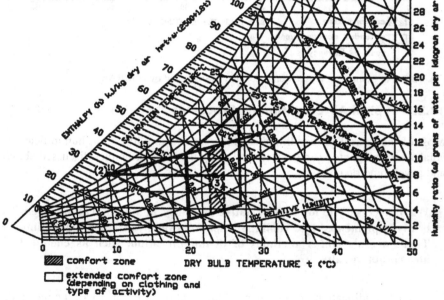

Fig. 13.4 Dehumidification by a cooling section followed by a re-heating section: Basic equipment and psychrometric chart representation

partial pressure and constant humidity ratio so that a new temperature (point 3) is reached with a relative humidity less than 100 %.

In order to avoid re-heating, which is an energy-consuming operation, dehumidification systems can bring the moist air mixture near the comfort zone instead of reaching saturation.

Humidification is implemented when it is necessary to increase the moisture content of the air.

It can be achieved by steam or by water injection. In the first case, the humidity ratio and the dry-bulb temperature will increase. In the latter the temperature at the outlet will be lower than at the inlet.

Cooling by evaporation can be done in hot dry climates. This is accomplished by spraying water into the air or by forcing contact with water. Because of the low humidity of the inlet air, part of the water evaporates, thus lowering air temperature and increasing relative humidity. If the inlet relative humidity value is small, the final humidity can be considered comfortable. No energy is added, so this phenomenon can be considered an adiabatic process (like humidification by water injection). As the constant enthalpy curves of the mixture are close to those of constant wet-bulb temperature, the process occurs at roughly constant wet-bulb temperature, as is shown by psychrometric charts.

Adiabatic mixing of two moist airstreams may occur when hot and cold streams are mixed or when exhaust air or external air is added.

13.3 Heating and Cooling Load Calculation

The design of HVAC systems requires a detailed analysis of cooling and heating loads. The choice among different systems depends on many factors such as building layout and occupancy, installation and maintenance requirements, capital and operating cost, energy cost, and production requirements.

The first step in designing an HVAC system is the calculation of heating and cooling loads.

To calculate the heating load, heat transmission losses through walls, glass, ceilings, floors, and other surfaces must be considered together with infiltration losses. Required indoor and outdoor temperatures must obviously be taken into account. In addition, the energy required to warm outdoor air to the space temperatures must be considered. The sum of all these terms represents the total heating load.

The internal heat sources (sensible and latent heat produced by people, motors, and machinery) may affect the size of the heating system and its operating modes. In industrial plants (manufacturing, data center, others) and commercial buildings, internal heat sources can supply a significant portion of the heating requirements.

Table 13.1 Typical loads from occupants in conditioned spaces

Type of activity	Total heat (W)	Sensible heat (W)	Latent heat (W)
Sleeping	70	50	20
Sitting quietly	110	75	35
Sitting light work	140	80	60
Standing light work	150	80	70
Light work in factory	250	100	150
Dancing	275	100	175
Heavy work in factory	450	180	270
Sports	600	250	350

In cooling calculation, the heat (sensible or latent heat) which enters or is generated within a space must be calculated. It can be due to transmission (as in heating mode) to solar radiation, to outdoor infiltration and ventilation (sensible and latent heat), to people (depending on the activity performed 100–400 W/person, of which latent heat represents 30–60 %: the greater the physical effort, the higher the value), and to other sources of sensible heat such as lighting, motors, and other equipment (with values depending on their power and efficiency).

Details of these calculations can be found in specialized handbooks and in ASHRAE publications. Table 13.1 lists typical loads from occupants.

A basic simplified relationship for a preliminary evaluation of heating and cooling loads can be formulated as follows:

$$Q = (c_d + 0.35 \times n_v) \times VB \times \Delta t \pm Q_i$$

where

Q = total load (W).

c_d = total building dispersion coefficient (W/m^3 K).

n_v = rate of ventilation in number of building volumes per hour (1/h).

$0.35 n_v$ = air heat capacity flow rate (W/m^3 K); only the contribution of the sensible heat is here considered (see enthalpy of the mixture in Sect. 13.2).

VB = volume of the building (m^3).

Δt = average temperature difference between outside and inside.

Q_i = internal source contribution (W) to be subtracted from heating consumption in cold seasons and to be added to cooling consumption in hot seasons. In cooling mode, the internal contribution from lighting and occupants reaches 10–20 % of the total load. Higher values can be found in particular situations.

Values of the above coefficients depend on the insulation of the building (see also Chap. 8), on the heat capacity (specific heat multiplied by the mass) of the building structures affected by indoor temperature, on climate, and on the ratio between the total dispersion surface (ground floor, ceiling, external surfaces with their windows) and the building's volume.

The colder the average outdoor temperature, the lower the total building coefficient c_d (typical values range from 0.35 to 1 W/m^3 K) must be; the higher the dispersion surface/volume ratio, the higher the c_d coefficient.

The n_v coefficient generally does not exceed unity if referred to the building gross volume, so that average values of $(c_d + 0.35n_v)$ ranging from 0.7 to 1.3 W/m^3 K can be assumed for the total load calculation. Notice that for particular situations, the n_v coefficient can be even higher than 10.

Latent heat can be introduced in the simplified relationship (Q) by using the difference between the total enthalpy of the outside and inside air-water mixture (see Sect. 13.2).

13.4 Typical HVAC Systems

In HVAC systems the heating mode is generally operated by means of boilers, heat pumps, and recovered heat; occasionally, depending on local conditions, electrically heated systems can be used.

HVAC systems for cooling are basically refrigeration systems and evaporation systems.

Refrigeration systems use compressors or absorption systems (see Chap. 12). The refrigerant evaporates through an evaporator that can be an air/refrigerant or water/refrigerant exchanger. If the exchanger is placed in the room to be cooled, air is cooled directly when it moves across the coils. Condensation can be by water (with cooling towers) or by air.

In evaporation HVAC systems, only pumps are required to move water inside the circuit and to force contact with air.

Heat pumps, as individual units or centralized plants, provide both heating and cooling (see Chap. 12).

Auxiliary equipment comprises AHUs (air-handling units), pumps and fans, and control systems.

HVAC systems can be classified as (1) all-air systems, (2) all-water systems, (3) air-water systems, and (4) individual units. Class (1) systems serve the conditioned space primarily by means of a network of ducts. Class (2) systems are based on a piping network which delivers energy as water to distributed units, typically fan coil units. Class (3) systems are a combination of the previous systems. Class (4) systems are based on independent equipment.

Figure 13.5 shows a basic scheme of an air-handling unit where the main energy uses (electricity for fans, cold and hot water, steam) are pointed out. Notice that the share among them depends on the operating parameters, indoor and outdoor. Electricity is always the most important part of AHU energy consumption.

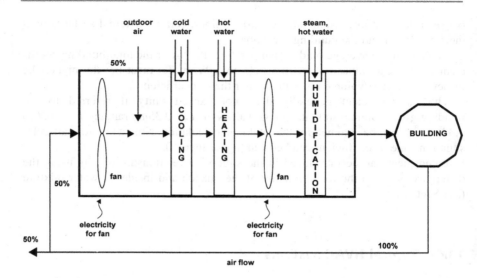

Fig. 13.5 A basic scheme of air-handling unit

13.4.1 All-Air Systems

All-air systems may be classified as follows: variable temperature and constant volume, constant temperature and variable volume, and variable volume and variable temperature.

Variable temperature and constant volume systems include single-zone systems, multi-zone with centralized blending, re-heating in each zone, dual-duct with mixing boxes in each zone.

Constant temperature and variable volume systems are mainly dual-duct systems, one duct with ventilation air in neutral conditions equal to those of the space to be conditioned and the other for cooling and heating.

Variable volume and variable temperature systems are mainly dual-duct with mixing boxes. Many combinations of the previous systems are possible.

Notice that in the re-heating mode, chilled air is supplied to each zone (for cooling mode) and then the air is re-heated according to the temperature required for each zone. This system consumes a great deal of energy because of the duplication of energy use. Variable air volume, which requires supply air at constant temperature, is based on valve control or on variable-speed fan control (by means of a local thermostat) of the airflow supplied to each zone, to ensure the supply only of the air needed by the real load of each zone. In this way energy consumption is reduced.

13.4.2 All-Water Systems

These systems use fan coils or unit ventilators, either of the under-the-window type or fixed at the ceiling, fed by piping network. For heating only, radiators are generally installed. In this case, heat release takes place through both convection (80–60 %) and radiation (20–40 %).

13.4.3 Air-Water Systems

In mixed systems, air and water, the energy supplied to each room is shared between water systems, such as radiators, fan coils, radiant systems (for floor or ceiling installation fed by steam or superheated water) or induction units, and the air supply system. There is a variety of combinations available.

The water system can supply most of the heating demand of the rooms while the main function of the air system is to meet air renewal and quality requirements and control humidity.

In induction systems, centrally conditioned primary air is supplied to the unit plenum at high pressure, which induces secondary air from the room to flow over the secondary coils where it is heated or cooled as the zone may demand. The so-called induction units are used in these systems.

13.4.4 Individual Units

These are package units for cooling. The condenser is generally air-cooled and expels heat from the room. A direct expansion cooling coil is usually installed. These units allow only temperature control.

13.4.5 Radiant Systems for Heating

Radiant energy can be used for heating by means of radiant tubes where hot fluids or hot gases flow. The temperature of hot gases is roughly 300–400 °C (572–752 °F). Exhausts from processes can conveniently be used.

This system can be economical if heating must be concentrated in defined areas inside the buildings, without heating the whole surface and volume. It avoids air stratification occurring in high-ceiling constructions.

Typical operating consumption of hot gas radiant systems for industrial applications is 0.02 Sm^3/h of natural gas per square meter of heated surface (175 W/m^2; 17.5 W/m^3 if building height is 10 m, 32.8 ft).

13.5   Centralized Control

Centralized control of HVAC systems allows significant savings on energy consumption and costs.

Control strategies for HVAC systems include many features such as night temperature setback for heating mode, night cooldown with outside air for cooling mode, outside and inside temperature control, minimization of outside air during heating, reduction of excessive ventilation, use of outside air for cooling when possible (free cooling in cooling mode), control of humidity, and implementation of cycles with night shutdown and morning warm-up.

Centralized control systems are based on the acquisition of a great number of parameters (many hundreds or thousands of values of temperature, humidity, and pressure) in all parts of the buildings and outside, and on the modeling of the buildings to make the proper control of each installation possible. Of course, off-line and online modeling programs are available to improve the design and the operation of HVAC systems.

In addition to the optimal control of temperature and humidity in each zone of the buildings, which is based on the availability of sensors and transducers to regulate the temperature of the air or the water supplied to the single zones, significant energy savings can be achieved by simply using outside air or by circulating exhaust air whenever possible.

The rate of heat recovery can be controlled on the basis of the dry-bulb temperature of the outside air or of its enthalpy value. The latter approach performs better, because it measures the heat content of outside air and avoids errors due to the difference between dry- and wet-bulb temperatures; the drawback is that it requires more maintenance.

13.6   Energy-Saving Systems

Energy saving in HVAC systems is primarily based on monitoring the existing equipment and on operating single components as efficiently as possible. Correct ventilation (an outside air supply plus any recirculated air that has been treated in AHU to maintain the desired air quality within a designated space) may allow significant energy saving (electric energy for fans, heating and cooling energy, and additional treatment, such as humidification and dehumidification, can be reduced).

The target must always be that of reducing loads by reducing losses from buildings and the negative contributions to cooling operation from machines and lighting; however, the required retrofitting is quite expensive.

Heat recovery from equipment and other sources such as exhaust air from buildings must always be practiced when possible.

Energy storage capability, such as chilled water or ice storage for cooling (see Sect. 4.9) and specially designed buildings with high thermal capacity structures (see Sect. 13.3), will improve the efficiency of the system by lowering the demand profile. Storage also makes it possible to shift energy consumption from peak hours to low-rate hours, thus increasing the energy cost saving.

When the ventilation rate is high, systems allowing heat recovery from exhaust air can be installed (see the example in Sect. 13.7).

Economizer systems are used to mix external air with return air; for instance in cooling mode return air at 24 °C (75 °F) can be mixed with cold outside air to maintain input air at 10–15 °C (50–60 °F) in a cooled space. If the outside temperature is around the preset value, 100 % outside air can be used. Thus the system works in the so-called free cooling mode. In hot seasons, the integration of outside air is kept as low as possible.

13.7 Practical Examples

Example 1 Heat recovery from exhaust air from buildings and inlet air entering buildings

Industrial laboratories, such as those in pharmaceutical and chemical factories, generally require high ventilation rates: 10–15 volumes of outside air which enters the building and must be heated or cooled to maintain the required temperature and relative humidity.

This example illustrates the energy saving by a group of exchangers (air to air with a water-glycol solution as intermediate fluid) which recovers the energy content of the exhaust outlet air in the cold season and of inlet air in the hot season.

The flow of the air without and with recovery equipment is represented in Figs. 13.6 and 13.7. Boiler-fed steam/air exchangers do all the heating. If recovery exchangers are installed, the system is modified as in the figures. There are two operation modes:

- Cold season mode (see Fig. 13.6):

  Exchangers are installed to recover sensible and sometimes also latent heat from the exhaust outlet air. The heat is released to a water-glycol solution, which circulates by means of pumps, and exchanges heat with the inlet air that otherwise would have to be heated by boiler-fed steam/air exchangers.

- Hot season mode (see Fig. 13.7):

  The inlet air passes through a cooling dehumidification section, followed by a re-heating section where boiler-fed steam/air exchangers are installed. The recovery system extracts heat from the total inlet air and transfers it to a part (40 %) of the air leaving the dehumidification section, thus reducing energy consumption for re-heating the latter. The total inlet air is cooled; thus, the consumption of the compressor refrigerating plant is reduced.

- Intermediate season mode:

  No recovery occurs and all heat is supplied by boiler-fed steam/air exchangers.

  The energy saving in cold season and hot season modes is calculated by considering only (1) the energy which the HVAC system (boiler plant and compressor refrigerating plant) would have required without recovered heat and (2) the additional consumption due to the new system. The calculations are summarized in Table 13.2.

Economic evaluation is shown in Table 19.4.

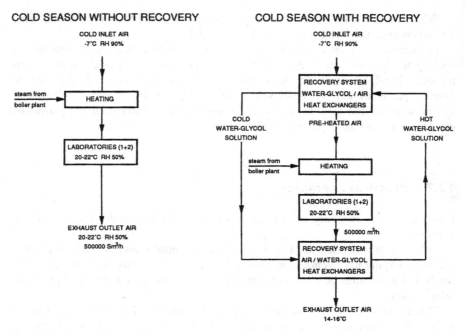

Fig. 13.6 Flow of the air without and with recovery equipment in the cold season

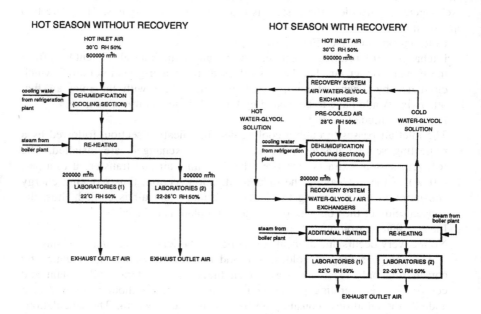

Fig. 13.7 Flow of the air without and with recovery equipment in the hot season

Table 13.2 Energy saving by heat recovery from exhaust and inlet air

	COLD SEASON MODE (see Fig. 13.6)		Recovery system	Equivalent fuel-fed boiler
a	Exhaust air flow rate RH 50 % 22 °C	m³/h	500,000	
$b = a/0.848$	Dry air mass flow rate	kg/h	589,623	
$c = b \times 0.00825$	Moisture flow rate	kg/h	4,864	
d	Inlet-outlet temperature increase	°C	6	
$e = d \times (1 + 0.00825 \times 1.8)$	Enthalpy drop per kg dry air	kJ/kg	6.1	
$f = e \times b$	Heat	kJ/h	3,590,271	3,590,271
$f' = f/3{,}600$		kW	997	997
$f'' = f/1.055$		Btu/h	3,403,101	3,403,101
g	Boiler efficiency			0.85
$h = f/g$	Boiler consumption (natural gas)	kJ/h		4,223,849
$h' = h/3{,}600$		kW		1,173
$h'' = h/1.055$		Btu/h		4,003,648
$i = h/34{,}325$		Sm³/h		123.1
l	Additional electricity consumption	kW	20	0
m	Working hours	h/year	3,750	3,750
	Energy consumption			
$n = l \times m$	Electric energy	kWh/year	75,000	0
$o = i \times m$	Thermal energy	Sm³/year	0	461,454

(continued)

Table 13.2 (continued)

	HOT SEASON MODE (see Fig. 13.6)		Recovery system	Equivalent fuel-fed boiler
A	After dehumidification RH 100 % 12 °C	m³/h	200,000	
$B = A/0.815$	Dry air mass flow rate	kg/h	245,399	
$C = B \times 0.0076$	Moisture flow rate	kg/h	1,865	
D	Temperature increase	°C	5	
$E = D \times (1 + 0.0076 \times 1.8)$	Enthalpy increase per kg dry air	kJ/kg	5.07	
$F = E \times B$	Heat	kJ/h	1,243,779	1,243,779
$F' = F/3{,}600$		kW	345	345
$F'' = F/1.055$		Btu/h	1,178,938	1,178,938
G	Boiler efficiency			0.85
$H = F/G$	Boiler consumption (natural gas)	kJ/h		1,463,270
$H' = H/3{,}600$		kW		406
$H'' = H/1.055$		Btu/h		1,386,985
$I = H/34{,}325$		Sm³/h		42.6
L	Additional electricity consumption	kW	20	0
M	Inlet air flow rate RH 50 % 30 °C	m³/h	500,000	
$N = M/0.87$	Dry air mass flow rate	kg/h	574,713	
$O = N \times 0.0133$	Moisture flow rate	kg/h	7,644	
P	Outlet–inlet temperature drop	°C	2.2	
$Q = P \times (1 + 0.013 \times 1.8)$	Enthalpy drop per kg dry air	kJ/kg	2.047	2.047
$R = Q \times N$	Heat extracted	kJ/h	1,176,943	1,176,943
$R' = R/3{,}600$		kW	327	327
$R'' = R/1.055$		Btu/h	1,114,997	1,114,997
S	Refrigeration plant COP			4.5
$T = R'/S$	Electric energy consumption	kW		72.6
V	Additional electricity consumption	kW	10	
Z	Working hours	h/year	1,250	1,250

Energy consumption

| | | | Cold + hot season modes | Equivalent boiler + compressor |
| | | | Recovery system | |
			(I)	(II)
$X = Z \times (L + T + V)$	Electric energy	kWh/year	37,500	90,766
$Y = Z \times I$	Thermal energy	Sm³/year	0	53,287

Energy consumption

			(I)	(II)
$EE = X + n$	Electric energy	kWh/year	112,500	90,766
$ET = Y + o$	Thermal energy	Sm³/year	0	514,741

Energy saving (II) − (I)

$SE = EE(II) - EE(I)$	Electric energy	kWh/year	−21,734
$ST = ET(II) - ET(I)$	Thermal energy	SDm³/year	514,741
$ST' = ST \times 34.325/41{,}860$		TOE/year	422

Cold season mode (design conditions)
Outdoor −7 °C, 19.4 °F, relative humidity 90 %
Inlet 22 °C, 71.6 °F, relative humidity 50 %

Hot season mode (design conditions)
Outdoor 32 °C, 89.6 °F, relative humidity 50 %
Inlet 22–26 °C, 71.6–78.8 °F, relative humidity 50 %

Cold season mode (average conditions)
Outdoor 2 °C, 35.6 °F, relative humidity 90 %
Inlet 22 °C, 71.6 °F, relative humidity 50 %

Hot season mode (average conditions)
Outdoor 30 °C, 86 °F, relative, humidity 50 %
Inlet 22–26 °C, 71.6–78.8 °F, relative humidity 50 %

Enthalpy (kJ/kg) per kg of dry air $= t + $ humidity ratio $\times (2{,}500 + 1.8 \times t)$ where t (°C)

Enthalpy (kJ/kg) increase (e) per kg dry air $= 6 \times (1 + 0.00825 \times 1.8)$

Enthalpy (kJ/kg) increase (E) per kg dry air $= 5 \times (1 + 0.0076 \times 1.8)$

Enthalpy (kJ/kg) drop (Q) per kg dry air $= 2 \times (1 + 0.013 \times 1.8)$

LHV of natural gas 34,325 kJ/Sm³

Example 2 Centralized control of HVAC systems

The centralized control of HVAC systems can be applied to industrial, commercial, and residential buildings in order to guarantee comfortable conditions with minimum energy consumption.

All kinds of software and hardware are available on the market. The estimate of energy saving would require knowledge of many details and data on the site concerned and on the hardware and software chosen.

As a general rule, a saving of 15 % on the previous energy consumption for HVAC can be assumed as an average value.

A building with 45,000 m^2 (485,000 ft^2) of floor and 100,000 m^3 (3,531 $\times$ 10^3 ft^3) of space to be heated and cooled is considered. The consumption is 600 TOE for heating in the cold season (2,500 h/year) and 1,600 MW h for cooling in the hot season (2,000 h/year). The energy saving associated with the installation of a centralized control system, based on 750–1,250 control points inside the building, is then estimated to be equal to 90 TOE/year and 240 MW h/year.

The economic evaluation is shown in Table 19.4 where only the contribution of energy saving is considered.

Other advantages, such as comfort and reduction of working hours for plant maintenance and operation, will certainly increase the economic advantages of a centralized control system (see also Sect. 17.8).

Facilities: Lighting

14

14.1 Introduction

Each building, each space in a factory, has its own characteristics and should be surveyed carefully in order to ensure the best quality and quantity of lighting. Depending on the kind of activity, energy consumption for lighting may be a significant part of the total energy consumption, like in supermarkets and commercial buildings (up to 20–30 %), or a very small percentage (less than 5 %) of the total energy consumption, like in factories.

Energy can be saved both by a proper choice of lighting systems for new installations or retrofitting and by a control of the energy flow (intensity control, on/off switching).

14.2 Basic Principles of Lighting Systems and Definitions

The basic concepts and definitions regarding lighting are set out briefly here below.
Luminous intensity of a lamp is expressed in candela (cd; see Table 2.1).
Lumen is the unit of luminous flux output from a lamp. The luminous flux is
  expressed in lumen (lm) = cd × sr (luminous intensity in candela multiplied by
  the solid angle in steradian).
Luminance is the amount of luminous flux per unit of area leaving the lamp surface
  in a particular direction. It is expressed in cd/m^2 or in cd/ft^2.
Illuminance is the amount of light falling on a surface. It is expressed in lux (lx) =
  lm/m^2 or in foot candle if a surface of 1 square foot is assumed (1 foot-
  candle = 10.764 lx).
Lamp efficiency is the ratio between the luminous flux from a lamp (lumen) and the
  electrical power consumed (watt): it is expressed as lm/W.
Reflectance is the ratio between the light reflected by a surface and the incident light
  on it.

G. Petrecca, *Energy Conversion and Management: Principles and Applications*,
DOI 10.1007/978-3-319-06560-1_14, © Springer International Publishing Switzerland 2014

Chromaticity or color temperature is an index of the light color that includes consideration of its dominant wavelength and purity. It is expressed in Kelvin temperature. The higher the temperature, the cooler the light (typical reference values are 4,100 K for the moon; 5,000 K for the sun; 10,000–25,000 K for blue sky). Chromaticity is defined as the temperature of a black body radiating light with the same color as the light source.

Color rendering or color rendition is an index of the effect of a light source on the color appearance of objects compared with color appearance under a reference light source (the value of color rendering for the reference light source equals 100).

Space distribution diagrams take into account that light emitted from a lamp or a luminaire may vary in luminous intensity depending on the direction. Diagrams are generally made on a plane section and report candle values referred to a flux of 1,000 lm to allow comparison among different lighting systems.

Luminaire (or fixture) is defined as a complete lighting unit consisting of a lamp or lamps and components designed to distribute the light (diffuser, reflector, lens, and others) and to connect the lamp to the power supply.

Coefficient of utilization (CU) or luminaire efficiency is the ratio of the luminous flux (lumen) reaching the working plane to the total flux (lumen) generated by the lamp. This coefficient takes into account light absorbed or reflected by walls, ceilings, and luminaries. Its values are generally found in charts in the manufacturer's catalogue and permit a quick estimation of the lighting that can be achieved with a given system.

Lamp and luminaire depreciation factor relates the initial illumination provided by a clean, new lamp and luminaire to the reduced illumination that they will provide at a particular point in time because of accumulated dirt and depreciation (see Fig. 14.1).

Task surface is the working area where the recommended levels of illumination specified by international associations should be applied. For the surrounding area the level of illuminance should be no more than 1/3 of the average level of the task surface. Table 14.1 shows typical values of illuminance for different industrial task surfaces; it ranges from 2,000 lx for high-precision operations to 50 lx for transit areas.

14.3 Typical Lighting Systems and Related Equipment

A lighting system may have many components: ballast, dimmer, lamp, diffuser, reflector, lens, and control equipment.

Systems are generally classified according to lamp categories, as follows: incandescent, fluorescent, high-intensity discharge such as mercury vapor, metal halide vapor, sodium vapor (high pressure and low pressure), and LED (light-emitting diode).

For all lamps, except for the incandescent lamps, the power factor is generally less than 0.5; to improve this value, capacitors must be provided (see Sect. 7.3). Tables 14.2, 14.3, 14.4, and 14.5 show the typical values of the main parameters.

Reflectance varies between 90 and 30 %; chromaticity between 2,500 and 3,000 K; and color rendering between 50 and 90.

As a general rule, with a required task surface illuminance ranging between 1,000 and 200 lx and a lamp efficiency ranging from 50 to 100 lm/W, typical values are 20–10 W/m^2. Of course, higher values can be found if incandescent lamps with less than 25 lm/W are used or if high levels of illumination are recommended for particular task surfaces.

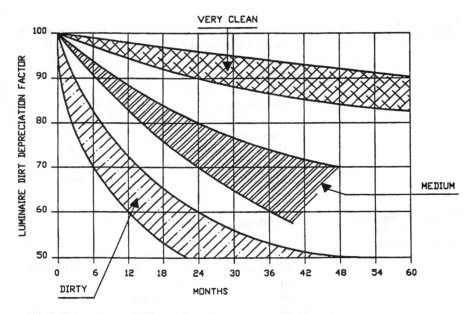

Fig. 14.1 Example of curves of the depreciation factor (ratio between the illumination provided at various points in time and that provided by the new system; it is expressed as a percentage)

Table 14.1 Recommended task surface illuminance

Task surface	Recommended illuminance	
	lx	fc
Very difficult task	1,000–4,000	93–372
Commercial surfaces	750–1,000	70–93
Material test, drawing, sewing, etc.	750–1,500	70–140
Office room	250–750	23–70
Transit area	50–150	5–15

Table 14.2 Average
values of reflectance
coefficients for different
materials and surfaces

Surface	Reflectance range (%)
Silver	92
Aluminum	70–60
Inox iron	65–55
White paint	90–70
White paper	80–70
White wall	90–70
Wood	30–20
Floor	60–20

Table 14.3 Chromaticity
of light sources

Source	Color temperature K
Daylight with blue sky	15,000–25,000
Overcast sky	5,000–15,000
Sunlight at noon	5,000
Early-morning sunlight	4,500
Sunlight at sunrise	1,700
Incandescent lamps	2,500–3,500
Fluorescent lamps	3,000–6,500
High-pressure sodium	2,000–2,500
High-pressure metal halide	3,500–4,500
LED	3,000–6,000

Table 14.4 Suggested
color rendering for
typical applications

Color rendering	Suggested applications
90	High-precision tasks, museum
80–90 warm tones	House, hospital, offices, school
80–90 cool tones	Textile, printing, mechanical engineering
60–80	Manufacturing industries
40–60	Heavy industries
20–40	In cases without particular requirements

14.3.1 Incandescent Lamps

Incandescent lamps consist of a resistive filament inside an evacuated glass bulb containing a trace of an inert gas. Electric current passing through the filament produces light. In a typical incandescent lamp about 5 % of the energy is converted into the visible spectrum, while the balance is radiated as infrared energy and heat. These lamps have a relatively short life (less than 2,000 h), because the filament gradually evaporates, reducing its diameter and increasing its resistance. This also reduces the light output to about 80 % of the rated lumens at the end of the life cycle. Incandescent lamps do not require ballast. Typical values are as follows: lamp efficiency 10–15 lm/W, color temperature 2,800 K, and color rendering 100.

Table 14.5 Typical lamp output characteristics

Type		Incandescent	Incandescent halogen	Fluorescent	Fluorescent high-frequency	Mercury vapor	Metal halide	High-pressure sodium	Low-pressure sodium	LED
Output characteristics										
Power (only lamp)	W	100	500	36	32	250	400	400	180	10
Lamp lumen	lm	1,500	12,500	2,600	3,200	14,000	32,000	48,000	33,000	1,000
Lamp efficiency	lm/W	15	25	72	100	56	80	120	183	100
Power (only ballast/dimmer)	W	0	0	10	4	16	25	35	30	<0.2
Lamp + ballast/dimmer efficiency	lm/W	15	25	57	89	52	75	110	157	98
Typical CU coefficient	%	70	70	70	70	40	40	40	40	90
Overall efficiency	lm/W	11	18	40	62	21	30	44	63	88
Life-span	h	1,000	3,000	7,500	12,500	10,000	10,000	10,000	10,000	50,000

Tungsten halogen lamps are also classified as incandescent lamps; the main operating characteristic is that the tungsten that evaporates and deposits on the bulb walls is removed and redeposited on the filament. They have many advantages such as maintaining the lumen output close to 100 % throughout life, higher color temperature (3,000 K), higher lamp efficacy (20–25 lm/W), and a life-span equal to 3,000–4,000 h.

14.3.2 Fluorescent Lamps

Fluorescent lamps require a ballast which provides an initial high-voltage pulse to start the discharge through a conducting vapor or gas (mercury, argon, krypton, others). The ballast also limits current flow through the lamp. Basically the ballast can be an inductor or a high-frequency electronic device. The latter has the advantages of (1) reducing power consumption of the lighting system by roughly 30 %, thus improving the luminaire efficiency, if used with fluorescent lamps designed to work at high frequency, tens of kHz instead of 50–60 Hz, and (2) dimming from 25 to 100 %.

The efficiency of the lamp increases with the arc length; this explains why fluorescent lamps are generally tubular. Typical values are 70–100 lm/W. The life-span is determined by the rate of loss of the electron-emitting material and it is also influenced by the number of times the lamps are switched on (typical life-span is 7,000–12,000 h instead of 1,000–2,000 h for incandescent lamps). The lamp depreciation factor reaches 90 % after 1,000 h and then slowly decreases to 85 % at the end of the life-span. The spectrum of the emitted light is modified by the phosphor coating on the inside of the tube.

Notice that roughly 20 % of input energy is converted to light and 40 % to infrared, while 40 % is dissipated as heat.

A choice of chromaticity and color rendering (generally 80–90) can be made depending on the phosphor mix of the lamp.

14.3.3 High-Intensity Discharge Lamps

High-intensity discharge lamps include mercury vapor, metal halide, and high-pressure sodium lamps. They work in the same way as fluorescent lamps and require starting devices (ballasts).

Mercury vapor lamps provide good color rendition (50–60), color temperature of 3,300–4,300 K, and good efficiency (35–60 lm/W). When the arc is extinguished, it cannot be switched on again until the vapor pressure is lowered to a point suitable for the applied voltage (3–7 min). They have a long life-span (up to 10,000 h) and a depreciation factor of more than 75 %.

Metal halide lamps have better efficiency (65–85 lm/W), better color rendition (65), and a color temperature of 4,500 K, thanks to the inclusion of metal additives. The life-span is generally 6,000–7,000 h with a depreciation factor of more than 75 %.

The high-pressure sodium is the most efficient high-intensity discharge lamp. Its golden-white color quality is a limiting factor for interior use, but improvement in color rendition (from 20 to 80) has made it acceptable also for interiors. The color temperature is around 2,000 K and efficiency is very high, up to 130 lm/W. The life-span is generally 10,000–15,000 h.

Low-pressure sodium lamps operate as discharge lamps. They need a special ballast. They have the highest efficiency up to 180 lm/W and higher but their monochromatic yellow light limits their use. The life-span is generally 10,000–15,000 h.

14.3.4 Light-Emitting Diode Lamps

An LED (light-emitting diode) lamp is a solid-state semiconductor device that converts electrical energy directly into light. Basically, it includes two regions: the p-region that contains positive electrical charges and the n-region with negative electrical charges. When voltage is applied and current begins to flow, the electrons move across the n-region into the p-region through the n-p junction by releasing energy. The dispersion of this energy produces photons with visible wavelengths. An LED driver is always required to convert input power into a current source which remains constant in spite of the voltage fluctuation; a dimmer regulates the input power from 20 to 100 %.

LED lamps have good efficiency (up to 100 lm/W and even a little higher), high color rendition (80–90), and a color temperature of 4,000 K and higher. The life-span is generally 25,000–50,000 h with a depreciation factor of 85 % at the end of the life.

14.3.5 Luminaires

With all kinds of lamps, luminaires play an important role in lighting systems.

The efficiency of luminaires is highly dependent on the reflectance of their interior surfaces (reflectors with silver films may reflect more than 90 % of the incident light), and exterior surfaces such as lenses (with polarizing materials and deflectors). Typical values range between 60 and 75 %.

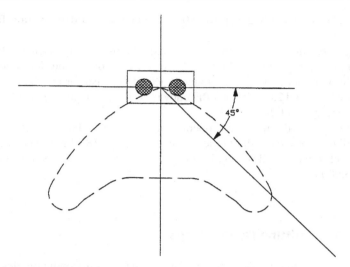

Fig. 14.2 Example of distribution diagram

Luminaire dirt depreciation is a major factor in maintaining good lighting levels at low energy cost. Typical depreciation factor curves are illustrated in Fig. 14.1 and they are provided by manufacturers. A good maintenance program may yield significant energy saving. In addition, other factors of depreciation such as room surface and furniture dirt depreciation, and aging and discoloration of luminaire shield materials, must be considered carefully in a lighting conservation program.

Luminaires must also avoid uncomfortably bright light (commonly called glare effect) on the task surface. Generally, light emitted within a cone with a 90° angle and with the axis along the vertical line from the luminaire to the floor does not contribute to direct glare. Inside the same cone loss of contrast on the task surface may occur; on the contrary, this effect is minimal outside the cone.

Shielding materials which direct light at particular angles can conveniently be used; distribution such as batwing distribution like in Fig. 14.2, where a high percentage of light is emitted near 45°, allows a good compromise among different requirements if the workstations are placed between rows of luminaires.

Lenses with polarizing materials and louvers (series of baffles generally arranged in a geometric pattern) are also used.

14.3.6 Lighting Control

The best way to save energy in lighting is to ensure that lights are turned on only when needed and are dimmed during periods when lower levels of lighting are required.

Single-program clocks, switches, photoelectric cells and dimmers, as well as a centralized computer-controlled switching system can be introduced, depending on the particular situation and working requirements.

Generally, the following rules must be remembered. Light must be turned on at the start of the business day, in specific areas for a watchman's tour of duty and at reduced intensity for cleaning crews. Light must be turned off at the close of the business day, on completion of cleaning, and in preset external sunlight conditions.

For more efficient control, each individual working area needs its own switch and large open areas need separate switches for different task zones. All these suggestions must be taken into account in designing new installations; for retrofitting, each case must be considered carefully because the cost of wiring may be too high.

Solid-state dimmers to vary the level between fully on and off give great flexibility, which can be valuable if the use of surfaces and sunlight condition change frequently. They reduce the voltage supplied to the lamp from a constant voltage source and save energy by reducing the light level to the right value for each situation.

Incandescent lamps can be totally dimmed from 100 % to zero. Fluorescent lamps and high-intensity discharge lamps cannot be dimmed totally because of their operation mode (at low voltage they cannot remain on). Specially designed high-frequency fluorescent and LED lamps can be regulated from 20 to 100 %.

14.4 How to Choose a Lighting System

The task-surface level must be created in accordance with local and international recommendations. Efficient lamps with high lumen, high efficiency (lm/W), and low lamp lumen depreciation should be used. Color temperature and color rendition must be checked in view of the kind of activity performed.

The choice of luminaires should be based on a high coefficient of utilization (CU) for the specific application and on the avoidance of depreciation due to dirt in the environment.

A preliminary evaluation of the lighting system can be made as follows:

(a) The illuminance ($lx = lm/m^2$) required at the task surface multiplied by the area (m^2) to be lighted is the total luminous flux required (lm).

(b) The lumen output for a single lamp multiplied by the coefficient of utilization of the luminaire (CU) gives the lumen output that will reach the task surface. The CU coefficient can be found in the manufacturer's charts where lamp and luminaire characteristics are taken into account together with the parameters of the surroundings (room dimensions, furniture, etc.). Typical values range between 35 and 60 %.

(c) The lumen output as at step (b) must be reduced by introducing the lamp and luminaire depreciation factors as given by the manufacturers (after a reference life period of the lighting system).

(d) The ratio between the value of the total luminous flux (lm) required as at step (a) and the value of lumen per lamp as at step (c) gives a preliminary evaluation of the number of lamps and luminaires required and then of the cost of the lighting system.

(e) The electric power required is then equal to the product of the single lamp power (or lamp with ballast, if necessary) multiplied by the number of lamps as at step (d). Knowing the working hours per year allows to calculate the energy consumption.

A more detailed knowledge of the lamp and luminaire space distribution diagram will place luminaires to ensure the task-surface level in all the working area while providing lower lighting levels in surrounding zones. In this way, capital and operating costs are reduced while the illumination requirements are satisfied.

As with any other investment in energy saving, economic analysis must take into account many aspects of the operating (energy and others) and capital costs.

A worksheet showing the main technical parameters together with capital and operating costs is given in Table 14.6 as a suggested guideline for evaluation.

14.5   Energy-Saving Opportunities

The foregoing considerations show that the energy-saving opportunities in lighting systems are as follows:

- A proper selection of lamps and luminaires to provide the task surface with the maximum efficiency (lm/W). This opportunity can be exploited both in new installations and in retrofitting.
- A planned maintenance program for replacing lamps, cleaning luminaires, replacing defective components, and cleaning surrounding surfaces in order to achieve the maximum exploitation at any time.
- A periodical checkup of the task surface.
- A control of lighting at a degree of complexity determined by the incidence of lighting costs on the energy bill: This control system may interact with other building or factory subsystems, in particular with the HVAC, because of the heat that lamps dissipate in surrounding areas.

14.6   Practical Examples

Example 1 Replacement of standard fluorescent lamps and luminaires by high-frequency ones (HF fluorescent lamps and electronic ballast).

In a shopping center with a shopping surface of 5,000 m^2 replace standard fluorescent lamps and luminaires with HF ones.

Table 14.6 Worksheet for energy-saving evaluation in lighting

			Reference	New system
Technical data				
a	Rated initial lumen per lamp	lm		
b	Input electric power per lamp	W		
$c = a/b$	Lamp efficiency	lm/W		
d	Input electric power per lamp + ballast (or dimmer)	W		
$e = a/d$	Lamps + ballast (or dimmer) efficiency	lm/W		
f	Coefficient of utilization—CU	%		
g	Lamp depreciation factor	%		
h	Dirt depreciation factor	%		
$i = e \times f \times g \times h/10^6$	Effective lumen per watt	lm/W		
$l = a \times f \times g \times h/10^6$	Effective lumen per lamp	lm		
m	Average lux required	lx		
n	Surface to be lighted	m^2		
$o = m \times n$	Total lumen required	lm		
$p = o/l$	Number of lamps	lamps		
$q = o/i = p \times d$	Total electric input power	W		
$KPI = q/n$	Installed power per unit of surface	W/m^2		
Capital cost				
r	Luminaire cost	U		
s	Auxiliary equipment cost	U		
t	Installation cost	U		
$u = r + s + t$	Total capital cost	U		
Operating cost				
v	Cost of replaced lamps	U/year		
x	Cost for replacing	U/year		
$y = v + x$	Total annual operating cost	U/year		
Energy saving				
W	Working hours	h/year		
$EC = W \times q/1{,}000$	Annual energy consumption	kWh/year		
$ES = EC(REF) - EC(NEW)$	Annual energy saving	kWh/year		
CA	Electric energy cost	U/kWh		
$CS = ES \times CA$	Annual energy cost saving	U/year		

The lighting system existing in the shopping center has 1,400 standard F96T12 lamps (input power 75 W per lamp) with two-lamp luminaires and standard ballasts of 10 W per lamp. The input power for each luminaire is 170 W. An average illuminance of 600 lx has been measured. The CU coefficient is 60 %.

The new lighting system includes two-lamp luminaires with HF 50 W input power per lamp and ballast input power of 6 W per lamp. The initial lamp efficiency is 100 lm/W; the lamp + ballast efficiency is 89.3 lm/W. The CU coefficient is still 60 %.

Table 14.7 Replacement of standard fluorescent lamps with HF lamps (Example 1)

			Reference (fluorescent)	New system (HF)
Technical data				
a	Rated initial lumen per lamp	lm		5,000
b	Input electric power per lamp	W	75[a]	50
$c = a/b$	Lamp efficiency	lm/W		100
d	Input electric power per lamp + ballast (or dimmer)	W	85[a]	56
$e = a/d$	Lamp + ballast (or dimmer) efficiency	lm/W		89.29
f	Coefficient of utilization—CU	%		60
g	Lamp depreciation factor	%		90
h	Dirt depreciation factor	%		90
$i = e \times f \times g \times h/10^6$	Effective lumen per watt	lm/W		43.39
$l = a \times f \times g \times h/10^6$	Effective lumen per lamp	lm		2,430
m	Average lux required	lx	600[a]	600
n	Surface to be lighted	m^2	5,000	5,000
$o = m \times n$	Total lumen required	lm	3,000,000	3,000,000
$p = o/l$	Number of lamps	lamps	1,400[a]	1,235
$q = o/i = p \times d$	Total electric input power	W	119,000	69,136
$KPI = q/n$	Installed power per unit of surface	W/m^2	23.8	13.8
Capital cost				
r	Luminaire cost	U		
s	Auxiliary equipment cost	U		
t	Installation cost	U		
$u = r + s + t$	Total capital cost	U		
Operating cost				
v	Cost of replaced lamps	U/year		
x	Cost for replacing	U/year		
$y = v + x$	Total annual operating cost	U/year		
Energy saving				
W	Working hours	h/year	5,000	5,000
$EC = W \times q/1,000$	Annual energy consumption	kWh/year	595,000	345,679
$ES = EC(REF) - EC(NEW)$	Annual energy saving	kWh/year		249,321
CA	Electric energy cost	U/kWh		
$CS = ES \times CA$	Annual energy cost saving	U/year		

[a]Basic input data for the existing system (reference)

The details of the energy-saving evaluation are shown in Table 14.7 which is derived from the worksheet of Table 14.6.

The annual energy saving is equal to 249,321 kWh.

See Table 19.4 for the economic evaluation.

Table 14.8 Comparison between HF fluorescent and LED lamps (Example 2)

			Reference (HF)	New system (LED)
Technical data				
a	Rated initial lumen per lamp	lm		3,000
b	Input electric power per lamp	W	50[a]	30
$c = a/b$	Lamp efficiency	lm/W	100	100
d	Input electric power per lamp + ballast (or dimmer)	W	56[a]	30.6
$e = a/d$	Lamp + ballast (or dimmer) efficiency	lm/W	89.29	98.04
f	Coefficient of utilization—CU	%	60	90
g	Lamp depreciation factor	%	90	90
h	Dirt depreciation factor	%	90	90
$i = e \times f \times g \times h/10^6$	Effective lumen per watt	lm/W	43.39	71.47
$l = a \times f \times g \times h/10^6$	Effective lumen per lamp	lm	2,430	2,187
m	Average lux required	lx	600[a]	600
n	Surface to be lighted	m^2	5,000	5,000
$o = m \times n$	Total lumen required	lm	3,000,000	3,000,000
$p = o/l$	Number of lamps	lamps	1,235[a]	1,372
$q = o/i = p \times d$	Total electric input power	W	69,136	41,975
$KPI = q/n$	Installed power per unit of surface	W/m^2	13.8	8.4
Capital cost				
r	Luminaire cost	U		
s	Auxiliary equipment cost	U		
t	Installation cost	U		
$u = r + s + t$	Total capital cost	U		
Operating cost				
v	Cost of replaced lamps	U/year		
x	Cost for replacing	U/year		
$y = v + x$	Total annual operating cost	U/year		
Energy saving				
W	Working hours	h/year	5,000	5,000
$EC = W \times q/1,000$	Annual energy consumption	kWh/year	345,679	209,877
$ES = EC(REF) - EC(NEW)$	Annual energy saving	kWh/year		135,802
CA	Electric energy cost	U/kWh		
$CS = ES \times CA$	Annual energy cost saving	U/year		

[a]Basic input data for the existing system (reference)

Example 2 Replacement of high-frequency lamps and luminaires (HF fluorescent lamps and electronic ballast) by LED ones.

In a shopping center with a shopping surface of 5,000 m^2 replace HF fluorescent lamps and luminaires with LED ones.

The lighting system existing in the shopping center has 1,235 HF lamps with 50 W input power per lamp and ballast input power of 6 W per lamp. The initial lamp efficiency is 100 lm/W; the lamp + ballast efficiency is 89.3 lm/W. The CU coefficient is 60 %.

The new lighting system includes 1,372 LED lamps and luminaires with 30 W input power per lamp and dimmer input power of 0.6 W (roughly 2 % of the lamp power) per lamp. The initial lamp efficiency is 100 lm/W; the lamp + dimmer efficiency is 98 lm/W. The CU coefficient is 90 %.

The details of the energy-saving evaluation are shown in Table 14.8 which is derived from the worksheet of Table 14.6.

The annual energy saving is equal to 135,802 kWh.

See Table 19.4 for the economic evaluation.

Heat Exchange and Recovery in Process and Facilities

15

15.1 Introduction

A great part of the total energy consumed all over the world is discharged as waste heat into the environment; part of this waste heat can be recovered by proper equipment if user demand matches it. The quality, defined mainly by temperature and quantity of waste heat and their profiles versus time, must match the user's demand, if energy recovery is to be significant from both the technical and economic points of view. Storage of waste heat is an expensive and not very effective approach with limited capability, and is little used (water and steam storage, refractory material storage systems, etc.—see Sect. 4.9).

Waste heat is available mainly as fluid, gas, or liquid streams at different temperatures. The streams can be more or less clean or corrosive, and this determines the possibility and cost of recovery. First, the temperature level qualifies the waste stream; the higher the temperature the greater the quantity of recoverable heat; ambient temperature is the minimum level of recoverable heat. Below or near this temperature, heat pumps can be used, but their application involves additional energy consumption, so their installation is not always convenient.

15.2 Opportunities for Heat Recovery from Process and Facility Plants

Table 15.1 lists the commonest sources of waste heat and their temperatures classified as high range (>925 K; >625 °C; >1,160 °F), medium range (900–500 K; 625–225 °C; 1,160–440 °F), and low range (500–300 K; 225–25 °C; 440–80 °F). The first two ranges belong to direct fired uses of fuel or other reactions and they represent a good source of energy still usable for cogeneration plants and for the direct replacement of process heat. In the higher temperature ranges attention must be paid to the equipment for heat recovery (proper material for high temperature, corrosion, etc.).

G. Petrecca, *Energy Conversion and Management: Principles and Applications*,   255
DOI 10.1007/978-3-319-06560-1_15, © Springer International Publishing Switzerland 2014

Table 15.1 Common sources of waste heat and their range of temperature

	K	°C	°F
Low-temperature range	*500–300*	*225–25*	*440–80*
Steam condensate	330–365	56–92	134–198
Cooling water from:			
Process machines	305–370	32–96	90–206
Air compressors	300–325	27–52	80–126
Internal combustion engines	340–395	66–122	152–252
Refrigeration condensers	305–315	32–42	90–107
Ovens and similar	350–505	76–232	170–450
Hot-processed materials	340–505	66–232	152–450

	K	°C	°F
Medium-temperature range	*900–500*	*625–225*	*1,160–440*
Steam boiler stack exhausts	500–750	226–375	440–890
Gas turbine exhausts	650–800	376–526	710–980
Internal engine exhausts	525–850	251–576	485–1,070
Furnace stack gases	700–900	426–625	800–1,160
Catalytic crackers	700–900	426–625	800–1,160

	K	°C	°F
High-temperature range	*>925*	*>625*	*>1,160*
Glass melting furnace	1,250–1,800	977–1,526	1,791–2,780
Cement kiln	900–1,000	626–726	1,160–1,340
Refining furnace	925–1,350	651–1,076	1,205–1,970
Steel heating furnace	1,200–1,300	928–1,026	1,702–1,880
Waste incinerator	925–1,250	652–977	1,206–1,791

The low-range heat sources, generally air and water cooling streams from process and facility equipment, are available for producing or preheating hot water.

> **The quantity of heat available in any stream can generally be quantified in terms of enthalpy (see Sect. 15.3):**
> $$Q = m \times h$$
> **where**
> **Q = total heat flow rate of the stream (W; Btu/h).**
> **m = mass flow rate of the stream (kg/s; lb/h).**
> **h = specific enthalpy of the stream (J/kg; Btu/lb).**

The mass flow rate can be calculated from the volume flow rate by introducing the density of the stream (m = mass density × volume flow rate).

It is important that every heat recovery system includes auxiliary equipment for heat discharge, generally into the environment, when waste heat demand from the user is interrupted or reduced. Otherwise, unacceptable operating conditions or interruptions would be imposed on the plant producing waste heat.

15.3   Heat Exchange and Recovery Systems

Heat exchangers are devices which transfer energy between fluids at different temperatures by heat transfer modes such as those discussed in Chap. 8.

Heat exchangers perform both heat recovery from waste heat and heat transfer from heating media to end users (see Sect. 8.4).

The exchange between two streams, a hot stream to be cooled and a cold stream to be heated, is induced either by mixing the two fluid streams to form a third stream at intermediate temperature or by a system which separates the two streams, thus allowing heat exchange through separating surfaces and sometimes by an intermediate fluid.

The first system, for example a vessel where hot and cold streams are mixed directly, has limited application due to the problems of contamination and of different pressures which often make it impossible to operate.

The second system is based on the use of heat exchangers often called closed heat exchangers or recuperators, available in many different types. Such stream characteristics as chemical composition, physical phase, temperature, and pressure affect the choice of heat exchangers and auxiliary equipment.

The following sections cover the basic principles of operation and the selection criteria for the commonest heat exchangers.

15.4   Basic Principles of Heat Exchanger Operation

The parameters which govern the functioning of a heat exchanger are the following:
- Maximum pressure: Notice that many exchanger types work only at low pressure; thus the choice among different types must be made carefully regard to this.
- Temperature range.
- Fluid compatibility both between fluids and construction materials and between fluid streams in the case of failure.
- Range and size available for each unit: Obviously, many units can be assembled together but this may involve economic and technical problems.

Table 15.2 lists the more frequently used types with observations on the main parameters listed above. All heat exchangers are classified mainly by flow arrangement parallel flow, counterflow, crossflow, and mixed flow (see Figs. 15.1 and 15.2).

In order to choose among the different types available, the same basic principles must be followed, from which a few suggested methods are derived.

Table 15.2 Typical parameters for common heat exchangers

Exchanger type	Max pressure		Range of temperature		Normal size (m²)	Notes
	MPa	psi	°C	°F		
Shell and tube	25–30	3,626–4,351	−200 to 600	−328 to 1,112	10–1,000	General purposes
Double pipe	>30 (shell) >150 (tube)	>4,351 (shell) 21,755 (tube)	−100 to 600	−148 to 1,112	0.25–200	No limitation
Gasketed plate	1–2	145–290	−25 to 175	−13 to 347	1–1,250	Limitations due to gaskets
Welded plate	3	435	>400	>752	>1,000	Differential pressure less than 3 MPa
Convection bank (gas/liquid)	0.1	14.5	<700	<1,292	<500	Usually for water heating
Gas to gas	0.1 (shell) >0.1 (tube)	14.5 (shell) >14.5 (tube)	250–400	482–752	1,000–3,000	For waste gas stream recovery
Air cooled	>0.1 (tube)	>14.5 (tube)	No limitation		5–200	Heat stream from process

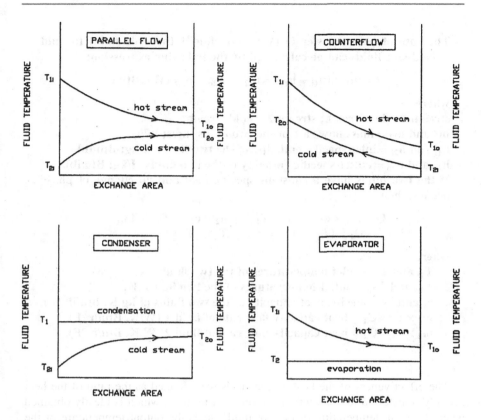

Fig. 15.1 Temperature profiles for different types of exchangers

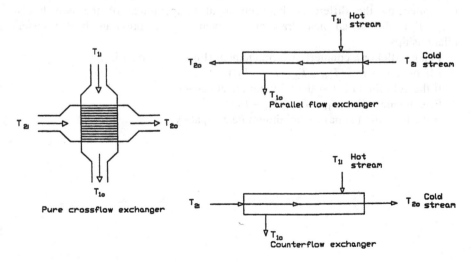

Fig. 15.2 Basic arrangements of different types of exchangers

The rate of heat transfer Q (W) from fluid 1 (hot-side fluid) to fluid 2 (cold-side fluid) can be calculated by the following expression:

$$Q = m_1 \times (h_{1i} - h_{1o}) = m_2 \times (h_{2o} - h_{2i}) \text{ (W) (Btu/h)}$$

where
stream 1 = hot stream; stream 2 = cold stream.
m_1 and m_2 = mass flow rate of the two fluids (kg/s; lb/h).
h_{1i} and h_{2i} = inlet-specific enthalpy of the two fluids (J/kg; Btu/lb).
h_{1o} and h_{2o} = outlet-specific enthalpy of the two fluids (J/kg; Btu/lb).
If the two fluids have a constant specific heat and no change of phase occurs, then

$$Q = m_1 \times c_1 \times (T_{1i} - T_{1o}) = m_2 \times c_2 \times (T_{2o} - T_{2i})$$
$$= C_1 \times (T_{1i} - T_{1o}) = C_2 \times (T_{2o} - T_{2i})$$

where
T_{1i} and T_{2i} = inlet temperatures of the two fluids (K, °C, °F).
T_{1o} and T_{2o} = outlet temperatures of the two fluids (K, °C, °F).
c_1 and c_2 = specific heat capacity of the two fluids (J/kg K, Btu/lb °F).
$m_1 \times c_1 = C_1$ = heat capacity flow rate of fluid 1 (W/K, Btu/h °F).
$m_2 \times c_2 = C_2$ = heat capacity flow rate of fluid 2 (W/K, Btu/h °F).

The effectiveness of the heat exchanger is then defined as the ratio of the heat actually transferred to potential maximum heat transfer, which is ideally obtained when the input temperature of the hot fluid equals the output temperature of the cold fluid ($T_{1i} = T_{2o}$; thus, $T_{2o} - T_{1o} = T_{1i} - T_{2i}$). The maximum heat transfer is equal to the minimum heat capacity flow rate of the two fluids ($m_1 \times c_1$ or $m_2 \times c_2$) multiplied by the difference between input temperatures of the two fluids ($T_{1i} - T_{2i}$); the actual heat transfer is given by the previous heat transfer relationships.

In a parallel or a counterflow arrangement (Figs. 15.1 and 15.2):
Effectiveness $E = (T_{2o} - T_{2i})/(T_{1i} - T_{21})$
If the cold fluid (2) has the minimum heat capacity.
Effectiveness $E = (T_{1i} - T_{1o})/(T_{1i} - T_{2i})$
If the hot fluid (1) has the minimum heat capacity.

A general expression of the heat transferred from one fluid to the other through the heat exchanger's separating surfaces can be formulated as follows (see also Sect. 8.3 for heat transfer modes):

$$Q = U \times A \times AT \ (W) \ (Btu/h)$$

where

$U =$ overall heat-transfer coefficient $= 1/R_{th}$ ($W/m^2 K$; Btu/h ft^2 °F).

$A =$ surface area of the exchanger, generally taken as the surface in contact with one fluid or stream (m^2; ft^2).

$\Delta T =$ effective temperature difference between the streams, assumed constant in any point of the exchanger surface, generally expressed as the log-mean difference between the temperatures of the fluids at the two sides of the exchanger. Notice that these differences are not the same for parallel and counterflow arrangements (K; °C; °F).

With a pure parallel flow exchanger, which is the simplest heat exchanger, and on the assumption of constant specific heats and constant overall coefficient U along the exchanger, ΔT must be introduced as follows (see Fig. 15.1 for fluid temperatures at both sides of the exchanger):

$$\Delta T = \Delta T_{LM} = \frac{(T_{1i} - T_{2i}) - (T_{1o} - T_{2o})}{\ln((T_{1i} - T_{2i})/(T_{1o} - T_{2o}))}$$

With a pure countercurrent flow exchanger, which is the most efficient heat exchanger, and on the assumption of constant specific heats and constant overall coefficient U along the exchanger, ΔT must be introduced as follows (see Fig. 15.1 for fluid temperatures at both sides of the exchanger):

$$\Delta T = \Delta T_{LM} = \frac{(T_{1i} - T_{2o}) - (T_{1o} - T_{2i})}{\ln((T_{1i} - T_{2o})/(T_{1o} - T_{2i}))}$$

In practice, many exchangers can be derived from a counterflow device, so that the effective temperature difference is defined as follows:

$$\Delta T = FT \times \Delta T_{LM} (\text{pure counterflow})$$

where

FT is a correction factor of the log-mean difference of temperature, generally obtainable from curves (see Fig. 15.3) where FT value is expressed as functions of R or P parameters:

$$R = (T_{1i} - T_{1o})/(T_{2o} - T_{2i}) = C_2/C_1$$

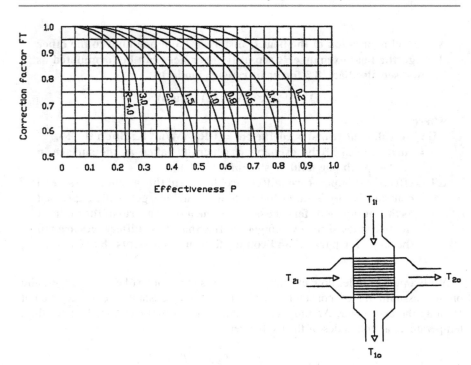

Fig. 15.3 Correction factor FT for a crossflow exchanger (P and R parameters are calculated on the assumption that the cold and hot fluids have the same heat capacity)

$$P = (T_{2o} - T_{2i})/(T_{1i} - T_{2i}) = (T_{1i} - T_{1o})/(T_{1i} - T_{2i})$$

The parameter R is the ratio between the difference of input/output temperatures of hot fluid (1) and cold fluid (2), that is, the ratio between heat capacity flow rates of cold fluid (2) and hot fluid (1). The parameter P is the effectiveness already defined on the same assumption as for parameter R, provided that the cold and hot fluids have the same heat capacity.

The overall heat-transfer coefficient U is not constant along the heat exchangers and depends on many factors such as the geometry of the exchanger, the stream speed, the temperatures of the streams, and phase changes. An exact calculation generally requires an iterative method and it is made by the manufacturer. For a preliminary evaluation, a constant value can conveniently be taken from reference tables that generally give a set of values. Table 15.3 reports some typical values of the U coefficient for a wide range of applications. Values of U for the commonest heat exchangers are also reported in Tables 15.4 and 15.5.

Table 15.3 Typical range of values of overall heat-transfer coefficients U

Application	Range of values of U coefficient	
	W/m^2 K	Btu/h ft^2 °F
Steam condenser	1,000–5,000	176–882
Water to water	500–2,000	88–353
Water to heavy organic fluid	100–400	18–70
Water to medium organic fluid	200–500	35–88
Water to light organic fluid	300–600	53–106
Steam to heavy organic fluid	60–300	11–53
Steam to medium organic fluid	300–400	53–70
Steam to light organic fluid	400–1,000	70–176
Water to air with finned tube	30–60	5–11
Steam to air with finned tube	30–300	5–53
	(a)	(b) = a × 0.1761

Notes: Heavy organic fluid: oil, etc. Medium organic fluid: gasoil, etc. Light organic fluid: Freon, etc.

Table 15.4 Shell and tube parameters ($Q/\Delta T$ ranges between 1,000 and 100,000 W/K)

Shell and tube

Cold-side operation	U coefficient	Hot-side operation			
		Gas (0.1 MPa) (14.5 psi)	Gas (1 MPa) (145 psi)	Process water	Condensing steam
Gas (0.1 MPa) (14.5 psi)	W/m^2 K (a)	55	93	102	107
	Btu/h ft^2 °F (b)	9.7	16.4	18	18.8
Gas (2 MPa) (290 psi)	W/m^2 K (a)	93	300	429	530
	Btu/h ft^2 °F (b)	16.4	52.8	75.5	93.3
Cooling water	W/m^2 K (a)	105	484	938	1,607
	Btu/h ft^2 °F (b)	18.5	85.2	165.2	283
Boiling water	W/m^2 K (a)	105	467	875	1,432
	Btu/h ft^2 °F (b)	18.5	82.2	154.1	252.2

Note: (b) = (a) × 0.1761

> **Attention must be paid to the surface condition of the exchanger. This varies during the operation because of (1) film deposits, such as oil films; (2) surface scaling due to the precipitation of solid compounds from solution; (3) corrosion of the surface; and (4) deposit of solid in liquid or gaseous streams. This phenomenon is called fouling and a fouling factor is introduced to take it into account. The fouling factor decreases with increased fouling and effectiveness drops correspondingly.**

In the case of heat exchange between many hot and cold streams inside the same process, a single composite curve of all hot streams and a single composite curve of

Table 15.5 Double-pipe finned tube parameters ($Q/\Delta T$ ranges between 5,000 and 100,000 W/K)

Double-pipe finned tube

Cold-side operation	U coefficient	Hot-side operation			
		Gas (0.1 MPa) (14.5 psi)	Gas (1 MPa) (145 psi)	Process water	Condensing steam
Gas (0.1 MPa)	W/m^2 K (a)	21	56	77	70
(14.5 psi)	Btu/h ft^2 °F (a) (b)	3.7	9.9	13.6	12.3
Gas (2 MPa)	W/m^2 K	56	97	181	182
(290 psi)	Btu/h ft^2 °F (b)	9.9	17.1	31.9	32.1
Cooling water	W/m^2 K (a)	87	238	325	534
	Btu/h ft^2 °F (b)	15.3	41.9	572	94
Boiling water	W/m^2 K (a)	84	216	284	511
	Btu/h ft^2 °F (b)	14.8	38	50	90

Notes: (b) = (a) × 0.1761
$Q/\Delta T$ range from 5,000 to 100,000 W/K and higher
The parameter cost per unit of $Q/\Delta T$ decreases as $Q/\Delta T$ value increases

all cold streams can be described in the temperature-heat content (Q) diagram. Heat exchanger networks can be designed or revamped by means of specialized techniques such as the pinch method to obtain the best energy exchange.

15.5 Preliminary Evaluation Criteria for Heat Exchangers

A preliminary evaluation of the heat exchanger surface can be made by simplified approaches such as the log-mean temperature method and the effectiveness-NTU method.

15.5.1 The Log-Mean Temperature Method

Preliminary sizing and pricing of heat exchangers are possible if the U coefficient is estimated, preferably by introducing also a fouling factor, and if inlet/outlet temperatures are known. The heat exchanger area may be calculated as

$$A = (Q/\Delta T) \times (1/U)$$

The expressions given above can be applied to different flow-arrangement types of exchanger: parallel flow, counterflow, crossflow, and mixed flow.

The ratio ($Q/\Delta T$) is an index of the heat exchanger duty and from this value a preliminary evaluation of the exchanger cost can be made by multiplying the area A by a cost per unit area, generally taken from the manufacturer's tables.

The procedure for evaluating various types of heat exchangers involves the following steps:

1. The heat flow rate Q of the stream is calculated using the relationships given above ($m \times \Delta h$).
2. The log-mean difference of temperatures ΔT_{LM} is calculated on the basis of the inlet/outlet temperatures and then corrected by factor FT.
3. The ratio $Q/\Delta T_{LM}$ should be calculated for each type of exchanger.
4. If U is estimated (see Tables 15.3 and 15.4), the exchanger surface is obtained and then the cost of the exchanger. Alternatively, curves or tables which give the cost per unit of W/K ($=$W/°C) or of W/°F corresponding to different values of U can conveniently be used to determine the cost of the exchanger.

15.5.2 Effectiveness-NTU Method

This approach is based on the introduction of a parameter defined as number of transfer units NTU $= A \times U/C_{min}$ where A is the effective heat transfer area, U is the effective overall heat-transfer coefficient, and C_{min} is the minimum heat capacity flow rate. The parameter NTU, which is an adimensional quantity, is indicative of the size of the heat exchanger.

The effectiveness of typical heat exchangers versus the parameter NTU is expressed by means of curves. Different curves can be described for different values of C_{min}/C_{max} (Figs. 15.4 and 15.5). These curves are plotted using the log-mean difference expression.

Notice that a linear relationship exists between effectiveness and the parameter NTU until the value of the latter reaches 1; after this point the curve begins to knee, which means that effectiveness can be increased with a corresponding higher increase of the NTU parameter. This involves paying more and more for exchangers with high effectiveness.

This method is very useful when the inlet and outlet temperatures in the heat exchanger are not known.

The following steps outline a preliminary design for a heat exchanger made by the NTU method, if only input temperatures and heat capacity flow rates of the fluids are known:

1. A value of the overall heat-transfer coefficient U is estimated.
2. The heat capacity flow rates of the two fluids are known ($m \times c$).
3. For a given effectiveness, which corresponds to a given output temperature of one fluid, the value of the NTU parameter is derived from curves like those in Figs. 15.4 and 15.5.
4. The exchange surface value can be calculated from the NTU expression (NTU $= A \times U/C_{min}$) and the unknown outlet temperature is derived from the effectiveness expression.

Alternatively

3'. For a given heat exchanger surface the NTU parameter is calculated.

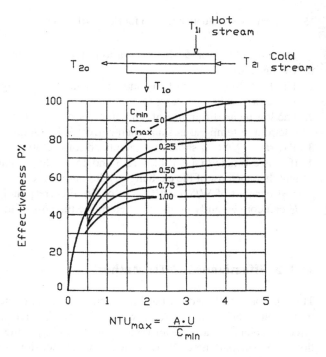

Fig. 15.4 Effectiveness versus NTU factor for a parallel flow exchanger

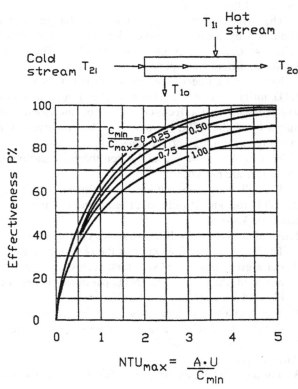

Fig. 15.5 Effectiveness versus NTU factor for a counterflow exchanger

4'. The effectiveness is derived from the same curves as before; then the actual heat flow rate is calculated (by means of the effectiveness definition), and the outlet temperatures derive from an energy balance.

15.6   A Classification of Heat Exchangers and Their Selection Criteria

The capital cost of a heat exchanger depends on many factors, such as
- Type of exchanger
- Tube diameter
- Number of passes
- Pressure
- Temperature
- Material
  Typical values of the main parameters are listed in Table 15.2.
  These factors can also be used to group heat exchangers as shown below.

15.6.1 Shell and Tube Heat Exchanger

The shell and tube heat exchanger is the most widely used in industry because of its flexibility both in operating parameters (temperatures, pressures) and in construction materials. It can be classified into fixed tube plate, U-tube type, or floating head. The exchangers are generally defined in terms of a standard nomenclature introduced by the Tubular Exchanger Manufacturers Association (TEMA) of the USA, based on the front-end, shell, and rear-end type.

U coefficient values vary from 50 to 2,000 W/m^2 K, from 9 to 352 Btu/h ft^2 °F (see also Table 15.4).

15.6.2 Double-Pipe Heat Exchanger

The double-pipe heat exchanger has the advantage of offering almost perfect countercurrent flow. It can be a single tube within an outer tube, or a group of tubes.

U coefficient values vary from 20 to 600 W/m^2 K, from 3.5 to 106 Btu/h ft^2 °F (see Table 15.5).

15.6.3 Gasketed-Plate Heat Exchanger

This exchanger is based on plates of stainless steel or titanium or other materials, which are packed together with gaskets. It can be enlarged at little cost by adding further plates, according to user requirements. Temperature and pressure limitations and the need for integrity of sealing are the major drawbacks of this type of exchanger.

U coefficient values vary from 300 W/m² K (53 Btu/h ft² °F) when high-viscosity liquid is on one side of the exchanger to 3,500–4,000 W/m² K (616–704 Btu/h ft² °F) with water or low-viscosity organic fluid.

15.6.4 Welded-Plate Exchanger

This exchanger, which has been designed to overcome the problems of the gasketed exchanger, has welded plates. If it is mounted within a shell, it allows achieving high pressures and high temperatures, equivalent to those in shell and tube exchangers. A large exchange area, more than 1,000 m² (10.764 ft²), can be exploited.

U coefficient values are similar to those of gasketed-plate heat exchangers.

15.6.5 Hot-Gas-to-Liquid Convective Bank System

This type of exchanger is used when waste heat from process is transferred to a liquid phase, generally water. It consists of groups of tubes with fins of cast iron or carbon steel or stainless steel. The working temperature ranges between 1,000 and 500 K (727–227 °C; 1,340–440 °F), depending on the location of the tubes, on their material, and on the gas characteristics. Attention must be paid to the condensation of acidic moisture.

U coefficient values vary from 15 to 30 W/m² K (2.6–5.3 Btu/h ft² °F), with water or other low-viscosity liquid on one side of the tube.

15.6.6 Gas-to-Gas Recuperative Heat Exchanger

This type of exchanger is widely used in industry in many applications such as the use of combustion gases to preheat combustion air, heat exchange in chemical reactors, and heat recovery from exhaust air in buildings. Many different configurations with tubes and plates and different materials are used (steel, cast iron, glass, plastic tube).

U coefficient values vary from 10 to 35 W/m² K (1.7–6.2 Btu/h ft² °F).

15.6.7 Heat Pipe Exchanger

The heat pipe exchanger uses an intermediate fluid to transfer heat from a hot gas to a cold fluid. The basic principle is as follows: heat from hot gases is transferred to the evaporation section. The intermediate fluid changes its state into vapor and moves along the heat pipe to the condensation section where it condenses by releasing heat to the cold stream to be heated. Water can be used as heat pipe fluid. Higher temperatures can be reached with organic fluid.

Large surfaces can be used for both sides.

A typical value of the *U* coefficient is 20 W/m² K (3.5 Btu/h ft² °F).

15.6.8 Rotary Regenerator

This exchanger consists of a disk, made of materials with a significant heat capacity, which rotates side by side between cold gas and hot gas ducts. The disk acts as an intermediate medium: it stores heat from the hot gas and it transfers heat when it moves in the area of the cold gas duct. The effectiveness of this system is very high, more than 90 %.

Attention must be paid to cross-contamination between the gas streams; this can be reduced by special devices such as labyrinth seals.

15.7 Practical Examples

Example 1 Cooling water-to-process water exchanger (log-mean method)

A process water flowing at 8 kg/s must be cooled from a temperature of 110 °C to a temperature of 40 °C using cooling water at the same flow rate. The inlet temperature of the cooling water is 15 °C and the desired outlet temperature is 85 °C. The pressure of the two fluids is around 0.3 MPa. A preliminary evaluation of possible heat exchangers is required (see Sect. 15.5.1 for the log-mean temperature method).

Given temperatures:

$$T_{1i} = 110 \text{ °C } T_{1o} = 40 \text{ °C}$$

$$T_{2i} = 15 \text{ °C } T_{2o} = 85 \text{ °C}$$

Step 1
Calculate the heat flow rate Q of the two streams:
– Stream 1 (hot water):

$$Q = 8 \times 4{,}186 \times (110 - 40) = 2{,}344{,}000 \text{ W} = 2{,}344 \text{ kW}$$

– Stream 2 (cold water):

$$Q = 8 \times 4{,}186 \times (85 - 15) = 2{,}344{,}000 \text{ W} = 2{,}344 \text{ kW}$$

Step 2
Calculate the difference between the temperatures of the fluids at the two sides of a counterflow exchanger:

$$T_{1i} - T_{2o} = 110 - 85 = 25 \text{ °C}$$

$$T_{1o} - T_{2i} = 40 - 5 = 25 \text{ °C}$$

$\Delta T = 25 \text{ K} = 25 \text{ °C}$ (in this case, it is constant along the counterflow exchanger).

$$R = (110 - 40)/(85 - 15) = 1$$

$$P = (85 - 15)/(110 - 15) = 0.73$$

Correction factor FT is preliminarily assumed to be equal to 1.

Step 3
Calculate $Q/\Delta T$ (W/K):

$$Q/\Delta T = 2{,}344{,}000/25 = 93{,}700 \text{ W/K}$$

Step 4
Calculate the exchange surface for different kinds of heat exchangers:
– From Table 15.4 (shell and tube exchanger):

$$U = 938 \text{ W/m}^2 \text{ K (typical value)}$$

$$\text{Area} = Q/(U \times \Delta T) = 2{,}344 \times 1{,}000/(938 \times 25) = 100 \text{ m}^2$$

– For plate exchanger U is assumed to be equal to 4,000 W/m^2 K

$$\text{Area} = Q/(U \times \Delta T) = 2{,}344 \times 1{,}000/(4{,}000 \times 25) = 23.4 \text{ m}^2$$

In both cases, if the parameter cost per unit of W/K is known, the cost of the heat exchangers can be found:

$$C \times (Q/\Delta T) = \text{cost of exchanger}$$

Although costs are not here included, the plate exchanger, which has the smaller surface, seems to be the more economical.

The energy saving is calculated with reference to a boiler plant which will heat the cold water flow from 15 to 85 °C as required from end users with an efficiency equal to 85 % (lower heating value as reference; 41,860 kJ/kg of oil).

In the case of 2,000 h/year recovery, it follows that

ENERGY   SAVING $= (2{,}344 - 3{,}600) \times 2{,}000/(41{,}860 \times 0.85) = 474{,}320$ kg$_{\text{oil}}$/
year $= 474.32$ TOE/year.

The economic evaluation is shown in Table 19.4.

Example 2 Cooling water-to-process water exchanger (effectiveness-NTU method)

The case is similar to that in Example 1; outlet temperatures of both fluids are supposed unknown (see effectiveness-NTU method in Sect. 15.4).
Given temperatures:

$$T_{1\text{i}} = 110 \text{ °C}$$

$$T_{2\text{i}} = 15 \text{ °C}$$

Step 1
Estimate a value of U coefficient:

$$U = 938 \text{ W/m}^2 \text{ K (shell and tube)}$$

Step 2
Calculate the heat capacity flow rate of the two fluids:

$$m_1 \times c_1 = 8 \times 4{,}186 = 33{,}488 \text{ W/K}$$

$$m_2 \times c_2 = 8 \times 4{,}186 = 33{,}488 \text{ W/K}$$

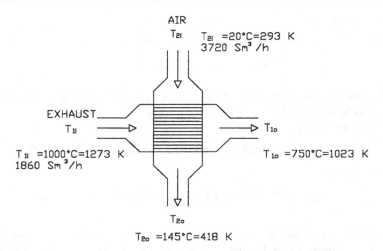

Fig. 15.6 Air exhaust gas crossflow exchanger. Operating data as for Example 2

Step 3
- The output temperature of fluid 1 is assumed to be $T_{1o} = 40\,°C$; the effectiveness is calculated as follows:

$$P = (110 - 40)/(110 - 15) = 0.736$$

- $C_{min}/C_{max} = 1$; from Fig. 15.5 it follows that $NTU_{max} = 2.9$

Step 4
Calculate the exchange surface and the outlet temperature of fluid 2:

$$NTU = A \times U/C_{min}.$$
$$A = NTU \times C_{min}/U = 2.9 \times 33{,}488/938 = 103\ m^2$$
$$P = (T_{2o} - 15)/(110 - 15) = (100 - 40)/(110 - 15) = 0.736$$
$$T_{2o} = P \times (110 - 15) + 15 = 0.736 \times 95 + 15 = 85\ °C$$

Example 2 has the same energy saving as Example 1.

Example 3 Air-to-exhaust stream exchanger (log-mean method)

A process requires 3,720 Sm^3/h of air at 145 °C, which can be obtained by heating air from the ambient temperature of 20 °C to the desired value by means of an exhaust gas stream. The exhaust is at 1,000 °C and has a flow rate equal to 1,860 Sm^3/h. The exchanger is a gas-to-gas type with a crossflow arrangement and a preliminary evaluation of its surface is required (see Sect. 15.4 for the log-mean temperature method).

Operating data at both sides of the heat exchanger are shown in Fig. 15.6.

Given temperatures:

$$T_{1i} = 1,000 \;^\circ\text{C} \; T_{1o} = 750 \;^\circ\text{C}$$

$$T_{2i} = 20 \;^\circ\text{C} \; T_{2o} = 145 \;^\circ\text{C}$$

Step 1

Calculate the heat flow rate of the two streams (specific heat of air and exhaust is assumed to be equal to 1 kJ/kg K)
- Stream 1 (exhaust):

$$Q = (1{,}860/3{,}600) \times 1.29 \times 1 \times (1{,}000 - 750) = 166 \text{ kW}$$

- Stream 2 (air):

$$Q = (3{,}720/3{,}600) \times 1.29 \times 1 \times (145 - 20) = 166 \text{ kW}$$

Step 2

Calculate ΔT by using the temperatures of the fluids at the two sides of the exchanger by means of the log-mean difference and the correction factor FT:

$$T_{1i} - T_{2o} = 1,000 - 145 = 855 \;^\circ\text{C}$$
$$T_{1o} - T_{2i} = 750 - 20 = 730 \;^\circ\text{C}$$
$$\Delta T_{LM} = 791 \text{ K} = 791 \;^\circ\text{C}$$
$$R = (1,000 - 750)/(145 - 20) = 2$$
$$P = (1,000 - 750)/(1,000 - 20) = 0.25$$
$$\text{FT} = 0.95 \text{ (see Fig. 15.3)}$$
$$\Delta T = \Delta T_{LM} \times \text{FT} = 791 \times 0.95 = 751 \text{ K} = 751 \;^\circ\text{C}$$

Step 3

Calculate $Q/\Delta T$ (W/K):

$$Q/\Delta T = 166 \times 1,000/751 = 221 \text{ (W/K)}$$

Step 4

Make a preliminary evaluation of a gas-to-gas exchanger surface and cost. U is assumed to be equal to 20 W/m^2 K; it follows that

$$\text{Area} = Q/(U \times \Delta T) = 166 \times 1,000/(20 \times 751) = 11 \text{ m}^2$$

If the parameter cost per unit of W/K is known, the cost of the heat exchangers can be found:

$$C \times (Q/\Delta T) = \text{cost of exchanger}$$

The energy saving is calculated with reference to a boiler plant which will heat the cold airstream (stream 2) as required by the process with efficiency equal to 85 % (lower heating value as reference; 41,860 kJ/kg of oil).

For 5,000 h/year recovery, it follows that

ENERGY SAVING $= (166 \times 3{,}600) \times 5{,}000/(41{,}860 \times 0.85) = 83{,}977 \text{ kg}_{\text{oil}}/\text{year}$
$= 84 \text{ TOE/year}$

The economic evaluation is shown in Table 19.3.

Waste and Energy Recovery

16

16.1 Introduction

Industry and human activities produce solid, liquid, and gaseous waste.

Industrial waste can be reused inside a factory as raw material and in energy-recovery plants. The main aspects to point out are the following: the chemical composition of waste, the potential reuse for internal process, the technology available for treatment, waste produced by treatment plants and the consequent air and water pollution, and also the economic viability, which should take environmental, energy, and operating costs into account.

Treatment of liquid and gaseous waste is widely practiced, to varying extents determined by local regulations, sensitivity to the problem and the kind of pollution concerned. Liquid and gaseous waste can be classified on the basis of pollutant content; specialized literature and practical experience are available to solve most cases either on-site or outside the factory. The energy content in liquid and gaseous emissions can be partially recovered by means of heat exchangers (see Chap. 15), to be used in process or in cogeneration plants (see Chap. 9), depending on the temperatures and pollutant content.

Solid waste from industry is classified in each country according to local regulations; nevertheless, the various systems have much in common. In an industrial country, like the example illustrated in Fig. 3.1, with a gross energy consumption of 160 MTOE/year (of which 50 MTOE/year is the share of industry), the amount of industrial waste is roughly 35 million t per year; it follows that, generally speaking, one can find 0.7–1 t of industrial waste per TOE of gross energy consumption in industry. In addition, notice that in a typical industrialized country each person disposes of about 1.5–2 kg/day (0.5–0.7 t/year) of urban waste; if all kinds of waste are considered, then in an industrial country such as the before-mentioned example with a population of 50 million, the total waste can reach 100 million tons, that is, 60 % in weight of the total TOE consumption.

G. Petrecca, *Energy Conversion and Management: Principles and Applications*,
DOI 10.1007/978-3-319-06560-1_16, © Springer International Publishing Switzerland 2014

Industrial solid waste is commonly broken down into four categories:
- **Inert waste** from extractive, brick and similar industries, to which can also be assimilated debris from building demolition. This can easily be reused for road construction and public works.
- **Urban waste**: This includes plastics, paper, rubber, and, in limited quantities, glass. These can be treated as urban waste.
- **Industrial waste** includes all industrial waste with the exceptions of the abovementioned categories (inert and urban waste) and hazardous waste. It involves some risks for the environment, but it can be made less harmful by appropriate treatment or reused in processes. It includes organic and inorganic materials, liquids, sludges, and solids with limited concentrations of pollutant.
- **Hazardous and infectious waste**, which is industrial waste with high concentrations of toxics or pollutants.

Urban waste is commonly landfilled, but this is no longer an acceptable disposal method because of environmental problems and the scarcity of space. Utilities are very interested in burning waste to produce electricity and steam, but air pollution must be avoided. Urban waste treatment is not discussed here in detail, our main concern being industrial waste and its on-site utilization.

Non-hazardous industrial waste can be treated as urban waste, but the quantity produced by a single medium-sized factory is generally too small for on-site reuse in heat-recovery plants. As a general rule, on-site treatment is economically viable if the quantity of waste is not less than hundreds of t/year (500–1,000). This depends on the type of waste and particular situations, but it can be assumed as a general indication.

16.2   Waste Management

Waste-management strategies comprise a few main sections, some of which are the same as in energy management (see Chap. 18): analysis of historical data, audits and accounting, analysis of local environmental regulations, engineering analyses, and investment proposals based on feasibility studies, personnel training, and information.

Several factors must be considered:
- Incidence of waste costs on turnover and added value
- Waste as a percentage of production
- Identification and quantification of waste following the flow of raw materials and energy inside the factory
- Selection of significant indexes valid for the whole factory and for single-production lines
- Possibility of recycling inside the factory or outside
- Level of pollution and compatibility with local regulations

Development of waste-management strategies at different levels depends on the importance of waste cost in comparison with other production factors and on local regulations, which may require following specific procedures, regardless of the waste quantity involved.

As a general rule, waste management is performed in three main ways (see also Fig. 4.5): (1) waste reduction by using clean technologies, (2) waste elimination, and (3) recycling of waste as raw material or energy.

The clean-technologies approach is the most appealing, but the most expensive and time consuming because it may require drastic transformations in all phases of the process, from raw-material input to packaging and delivery. Technologists must be involved in process change. Also important and not to be overlooked are the effects of the packaging phase on urban and industrial waste, due to the plastic and wood waste produced. In addition, the use of combustibles such as natural gas, liquid, and solid fuels with low sulfur concentrations (less than 1 % in weight) can be included in this area of waste management; it can be called clean technology in facilities, because gaseous pollutants from combustion are reduced without abatement plants at the end of the combustion process.

Elimination of solid and liquid waste by employing external agencies is the simplest but generally the most expensive policy for an individual factory. Quite often it is the only feasible one because the quantity of waste produced is too small to justify an independent plant. Nevertheless, local regulations may require storage and transportation in accordance with very strict rules, which may increase the final cost of disposal. It must be pointed out that the difficulties in finding suitable sites for this operation are growing dramatically, particularly in crowded areas such as the most industrialized part of the world.

In the case of liquid waste, water discharge to external agencies costs according to the pollutant content. Otherwise, liquid waste could be treated before discharge by means of mechanical, chemical, and aerobic systems, which consume energy.

In a similar way, the treatment of gaseous streams at high temperature must be performed in the factory, depending on local regulations (by burning the pollutant streams and by keeping them for a given time at temperatures higher than 900–1,000 °C (1,650–1,832 °F) and then cooling them to a lower temperature before discharge from the stack). Notice that the cooling phase can be economically performed by recovering heat from the exhaust through gas-to-gas exchangers, thus reducing the energy consumed in treatment.

Whenever possible, recycling as raw material or energy is always the most attractive approach.

The use of waste as raw material in the process itself remains the best form of waste management, because in this case waste does not have to leave the factory boundary. The use of waste coming from other factories and processes can be equally successful in a wider waste-management approach. It is a method practiced in foundries and cement industries by using plastics, rubber, liquid waste, and RDF; it requires separation of material flows in order to facilitate the recycling, whenever possible, at different steps of the process.

Although it is not always practicable because of environmental pollution problems and the corrosion of equipment by substances such as polyvinyl chloride (see combustion principles in Sect. 6.11), incineration of waste (RDF, sludge, and others) inside the factory itself is always an attractive possibility.

Pyrolysis, distillation, and anaerobic treatment of solid waste, sludges, and liquids produce a liquid or a gas suitable for use as fuel. Many of these treatments, such as anaerobic digestion processes, although often mentioned as economical treatments, are mostly confined to small plants. Energy recovery from these plants is low and the amount of recovered energy often only equals the input energy for auxiliaries. Nevertheless, unlike other energy-consuming systems as have already been mentioned, these treatments allow some energy saving.

16.3   Energy Recovery from Waste

Energy recovery from waste can be classified on the basis of the kinds of treatment and of waste available.

Heat recovery from hot liquid and gaseous streams by means of heat exchangers has been discussed in Chap. 15. It is the simplest and most economical method of recovery, but requires that the recoverable stream match the demand profile of the end user in quantity and in quality. In addition, the pollutant content of the stream, which might provoke corrosion, must be considered carefully. ,

Many types of solid and liquid waste can be stored when they are produced and then transformed into energy as liquid, solid, and gaseous fuels according to the end user demand.

Table 16.1 lists the main treatment technologies for energy recovery and a few basic operating data to be considered in a preliminary evaluation. These systems are subject to the following considerations:

- Urban and industrial solid waste can be burned in incinerators to produce hot water, steam, and electricity. So can industrial and urban sludges. Solid waste can be prepared as RDF (refuse-derived fuel) or as solid fuel pellets. Net heating values range from 16,000 to 21,000 kJ/kg.
- Liquid and solid waste can be treated by means of anaerobic digestion to produce biogas. The main parameter to be considered for liquid waste is the COD (chemical oxygen demand) content to which the production of biogas is generally referred.
- Other systems, such as pyrolysis, distillation, and gasification, are possible in particular applications, which must be investigated individually and evaluated carefully from the economic and technical points of view.

Table 16.1 Main technologies for energy recovery in waste treatment

Type of technology	Energy recovery		Note
Incineration	*Lower Heating Value (kJ/kg)*		
RDF fluff	16,000–21,000		
RDF pellet	16,000–21,000		
Urban solid waste	4,000–9,500		
Industrial solid waste	20,000–40,000		Depends on the type of refuse
Sludges	10,000		Moisture content 5 %
Anaerobic digestion (biogas)	*Lower Heating Value (kJ/Sm³)*	*Production (Sm³/kg)*	
Urban solid waste (landfilled)	18,000–22,000	0.35–0.45	Per kg of dry mass
Sludges (urban, agricultural)	15,000–20,000	0.6–1.5	Per kg of solid substance gasified
Liquid waste	20,000–25,000	0.2–0.4	Per kg of COD (liquid chemical oxygen demand); COD >5,000; methane 65 %
Biomass	15,000–20,000	Variable	Depends on the type of biomass

16.4   Waste and Energy Management Functions in Industry

Waste management is a strictly energy-related function, which concerns all factories and the activities of every person at work and at home.

If energy must be considered with an eye to the limited amount available and the search for new sources must always be continued, waste must be viewed in the perspective of its huge presence everywhere in the world. On the other hand, waste means energy: (1) energy can be produced from disposal at the end of a production process, and (2) energy is consumed both during the transformation of raw materials into the end product and during waste treatment.

Cost of waste treatment is certainly one of the factors to be considered in the economic analysis of investments (see Chap. 19).

Energy Management Strategies for Control and Planning

17

17.1 Introduction

Energy management strategies traditionally include six main sections: analysis of historical data; check of energy supply contracts; energy audit and accounting; evaluation of Key Performance Indicators (KPIs); engineering analysis, investment proposals and plant control based on feasibility studies; personnel training and information. Several factors must be considered:

- Incidence of energy costs on turnover and added value;
- Percentages of thermal and electric energy both as used in process and as primary energy;
- Identification and quantification of thermal and electric energy flows from transformation plants to users;
- Selection of significant Key Performance Indicators valid for the whole site, for facility plants and for every production line;
- Possibilities of heat recovery and centralized control of process and facility plants;
- Comparative incidence of energy used for process and services on total consumption.

> **Energy strategies can be implemented at different levels depending on the importance of energy in comparison with other production factors. Of course, also the absolute cost of energy consumed in the site is a parameter to be considered before starting any energy saving program.**

Energy strategies should be tuned up with waste strategies because of their close interdependence.

G. Petrecca, *Energy Conversion and Management: Principles and Applications*, 279
DOI 10.1007/978-3-319-06560-1_17, © Springer International Publishing Switzerland 2014

17.2   The Different Steps of an Energy Management Approach

Figure 17.1 shows the suggested basic approach to energy management. To set up an energy management program, it is necessary to know the organization and functioning of the site, factory and building.

If the energy management problem has not been treated before, all the sections shown in Fig. 17.1 must be covered. In the opposite case, each section of the program can be developed separately.

Energy analysis and audit include six main sections, summarized in Fig. 17.1 and detailed in the following:

1. *Analysis of historical data* regarding at least the last 3 years, if no significant changes have occurred in this period. For the unit cost of energy see also Sect. 19.4. The data can be organized on a monthly basis:

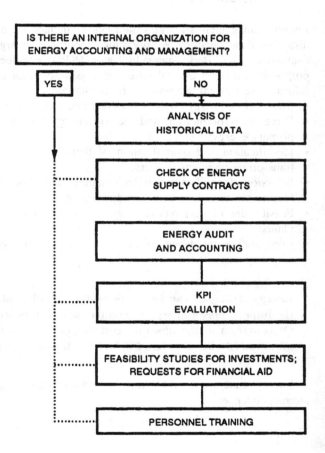

Fig. 17.1 A suggested basic approach to energy management: the program can be started at different levels

- Electrical energy consumption in kWh and power consumed in kW during different rate hours, if rate hours tariff is applied, and related costs. When different contracts exist, several tables can be used to report data referring to a single contract. The cost related to power factor can also be reported if applied.
- Natural gas consumption in Sm^3 or kWh and related costs.
- Oil consumption in kg and related costs.
- Gasoil consumption in liter (L) and related costs.
- LPG consumption in liter (L) and related costs.
- Coal consumption in kg and related costs.
- Other combustibles and related costs.
- Heat (steam, hot water) and related cost in case of district heating or other.
- Water consumption from both utilities and site wells in m^3 and related costs.
- Production grouped in main categories (in this example: productions x, y) expressed in t or number of pieces or some other significant unit. A unifying unit must also be employed in order to have a measure of the total production; otherwise, the working hours can be assumed as an index of the total production or the number of occupants or people entering the buildings.
- Waste production grouped in main categories expressed in t or in m^3. Energy content can be detailed if significant.
- SI units are used.

2. *Checking of energy supply contracts.* This is a preliminary check based on the analysis of the contracts with utilities in order to reduce costs without any significant change in the process and service operation modes. Depending on the local tariffication, the suitability of the contract for the demand profile (kilowatt load profiles with 15-min or 30-min integrated mean values, kilowatt-hour energy consumption during peak and off-peak hours, and load factor) must be investigated carefully in order to obtain the cheapest available rate.

   A more detailed analysis can be made when the energy situation of the factory is completely known.

3. *Energy audit and accounting*
   (a) Construction of site's energy and mass flow charts from input to output. The energy flow charts of process and facilities must be correlated with each other for a preliminary test of congruence. They must be congruent with site models as defined at step 3c.
   (b) Elaboration of data collected in step 1 and correlation among them in order to extrapolate significant parameters for the energy assessment of the site (from Tables 17.1, 17.2, and 17.3). The annual data used as reference year are derived from the monthly based data as in step 1:
      - Synthesis of energy consumption and costs as in Table 17.3. Percentage distribution of different forms of energy.
      - Yearly variations of energy consumption and related costs.
      - Yearly variations of different types of production.
      - Site or group of sites.
      - List of facility plants with rated power (Table 17.4).

Table 17.1 General
data XXXX–YYYY

Company name	XXXX
Location	YYYY
Gross covered area—production, warehouse, lab	10,000 m^2
Gross covered volume—production, warehouse, lab	50,000 m^3
Area and volumes with HVAC	8,000 m^2/40,000 m^3
Working hours/day	24 h/day
Area of activity	Food production
Production	10,000 t/year
Number of daily attendance (employees)	100 unit/day
Number of daily attendance (external)	10 unit/day

Table 17.2 Energy
consumption
XXXX–YYYY

Electric energy consumption (year . . .)	10,007,239 kWh/year
Electric energy reference cost (year . . .)	0.1 U/kWh
Supply voltage	11 kV
Natural gas consumption (year . . .)	1,805,339 Sm^3/year
Natural gas reference cost (year . . .)	0.25 U/Sm^3
Water consumption (year . . .)	35,916 m^3/year
Water reference cost (year . . .)	1 U/m^3
Total cost (electric + natural gas + water)	1,487,975 U/year

Table 17.3 Share of
energy and water
cost—XXXX-YYYY

Electric energy	1,000,724 U/year
Natural gas	451,335 U/year
Water	35,916 U/year
Total cost	1,487,975 U/year

(c) Modeling of the site: electric model, thermal model, water model, others as shown in Tables 17.5, 17.6, and 17.7.

The approach for the models is:
- List of users with their installed power or rated consumption per hour (electric process users, heat users, water and others like cooling, HVAC, compressed air).
- Percentage load factor depending on the operation mode in the reference period. Since this parameter is multiplied by the installed power to obtain the consumed power, the efficiency of the system (motors, distribution lines, etc.) is included in the percentage load factor.
- Consumed power equal to installed power multiplied by percentage load factor.
- Working hours per day, working days per year, and working hours per year.

Table 17.4 Facilities—XXXX–YYYY

Compressor cooling plant	Fluid and working temperature (°C)	Cooling power (kW)	Electric power (kW)	Rated COP
Chiller 1, 2, 3	Water 8 °C	2,400	600	4
Total		2,400	600	4
Chilled water pumps (10 pumps, 15 kW each)			150	

Cooling tower water pumps and fans	Fluid	Pump electric power(kW)	Fan electric power (kW)	
Pump tower no 1	Water	8	10	
Pump tower no 2	Water	15	20	
Pump tower no 3	Water	6	10	
Pump tower no 4	Water	18	20	
Total		47	60	

Compressed air plant	Pressure (MPa)	Electric power (kW)	Inlet air flow (Sm³/h)	kWh/Sm³
Air compressor no 1	0.70	150	1,378	0.11
Air compressor no 2	0.70	110	1,164	0.09
Total		260	2,542	0.10

Vacuum plant	Pressure (MPa)	Electric power (kW)		
Vacuum pumps (3 pumps, 10 kW each)		30		
Total		30		

HVAC–UTA	Rated air flow (m³/h)	Electric power (kW)	Zone identification	
Factory zone a	60,000	15		
Factory zone b	15,000	10		
Factory zone c	17,000	10		
Factory zone d	60,000	15		
Factory zone e	20,000	10		
Factory zone f	4,000	5		
Total	176,000	65		

Lighting	Number of lamps	Electric power (kW)	Surface (m²)	W/m²
Fluorescent lamps 36 W (indoor)	2,000	72	10,000	7.2
Discharge lamps (outdoor)	20	4.0	2,000	2.0
Total	2,020	76		

Boiler plant	Thermal power (kW)	Fluid	Input fuel	
Boiler 1, 2, 3	1,500	Steam	Natural gas	
Boiler 4	1,500	Hot water 60 °C	Natural gas	
Total	3,000			

Other equipment				
–				
–				

All values are rated value

Table 17.5 Electrical model—XXXXX–YYYY

User	Activity	Area	Product	Use	Rated power kW (kW)	Installed units (Unit)	Total rated power (kW)	Load factor/ efficiency (%)	Average input power (kW)	Daily working hours (h/day)	Working days (day/year)	Yearly working hours (h/year)	Input energy (kWh/year)	Input energy share (%)
Fluorescent lamps	Facility	Indoor	x, y	Inlight	0.036	2,000	72	70	50.4	24	312	7,488	377,395.2	3.77
Discharge lamps	Facility	Outdoor	x, y	Outlight	0.2	20	4	70	2.8	12	260	3,120	8,736	0.09
Chiller no 1,2,3	Facility	Outdoor	x, y	Chiller	200	3	600	50	300	24	365	8,760	2,628,000	26.26
Cooling tower no 1	Facility	Outdoor	x	Pump and fan	18	4	47	30	14.1	24	365	8,760	123,516	1.23
Cooling tower no 2	Facility	Outdoor	x	Pump and fan	35	1	35	30	10.5	24	365	8,760	91,980	0.92
Cooling tower no 3	Facility	Outdoor	y	Pump and fan	16	1	16	30	4.8	24	365	8,760	42,048	0.42
Cooling tower no 4 pumps and fans	Facility	Outdoor	y	Pump and fan	38	1	38	30	11.4	24	365	8,760	99,864	1.00
Chilled water pumps	Facility	Indoor	x, y	Pump	15	10	150	60	90	24	365	8,760	788,400	7.88

Compressor no 1	Facility	Indoor	x	ACO	150	1	150	50	75	24	365	8,760	657,000	6.57
Compressor no 2	Facility	Indoor	y	ACO	110	1	110	50	55	24	365	8,760	481,800	4.81
Vacuum pumps	Facility	Indoor	x	Vacuum	10	3	30	50	15	24	365	8,760	131,400	1.31
HVAC	Facility	Production x	x	AHU	15	1	15	40	6	24	365	8,760	52,560	0.53
HVAC	Facility	Production x	x	AHU	10	1	10	40	4	24	365	8,760	35,040	0.35
HVAC	Facility	Production y	y	AHU	10	1	10	60	6	24	365	8,760	52,560	0.53
HVAC	Facility	Production y	y	AHU	15	1	15	60	9	24	365	8,760	78,840	0.79
HVAC	Facility	Production y	y	AHU	10	1	10	60	6	24	365	8,760	52,560	0.53
HVAC	Facility	Production y	y	AHU	5	1	5	60	3	24	365	8,760	26,280	0.26
Process x	Process	Production x	x	Proc	500	1	500	35	175	24	365	8,760	1,533,000	15.32
Process y	Process	Production y	y	Proc	600	1	600	45	270	24	365	8,760	2,365,200	23.63
Other equipment	Facility	Indoor	x, y	Proc	50	1	50	45	22.5	24	365	8,760	197,100	1.97
Losses	Facility	Indoor	x, y	Loss	20	1	60	35	21	24	365	8,760	183,960	1.84
Total							2,527		1,151.5				10,007,239	100

Table 17.6 Natural gas model—XXXX–YYYY

Users	Activity	Area	Product	Use	Rated power (kW)	Load factor (%)	Efficiency LHV (%)	Average input power (kW)	Daily working hours (h/day)	Working days (day/year)	Yearly working hours (h/year)	Natural gas consumption (Sm³/year)	Output energy (MWh/year)	Share (%)
Boiler 1	Facility	indoor	x	Steam	500	75	90	416.67	24	365	8,760	381,422	3,404	21.13
Boiler 2	Boiler	facility	indoor	y	Steam	500	50	90	277.78	24	365	8,760	254,281	2,269	14.08
Boiler 3	Facility	Indoor	y	Steam	500	50	90	277.78	24	365	8,760	2,542,81	2,269	14.08
Boiler 4	Facility	Indoor	x, y	Hot water	1,500	60	90	1,000	24	365	8,760	915,413	8,170	50.70
Total					3,000			1,972.22				1,805,399	16,114	100

NHV natural gas 34,450 (KJ/Sm³)

NHV natural gas 1,036 (Btu/Sft³)

Table 17.7 Water model—XXXX–YYYY

Users	Activity	Area	Product	Use	Rated flow (m³/h)	Load factor (%)	Average flow (m³/h)	Daily working hours (h/day)	Working days (day/year)	Yearly working hours (h/year)	Water consumption (m³/year)	Share (%)
Cooling towers	Facility	Outdoor	x, y	Cooling tower	6.00	55	3.3	24	365	8,760	28,908	80.49
Boiler feed water	Facility	Outdoor	x, y	Boiler	0.5	100	0.5	24	365	8,760	4,380	12.20
Production	Process	Indoor	x, y	Process	0.1	100	0.1	24	365	8,760	876	2.44
Drinkable water	Facility	Indoor	x, y	Sanitary water	0.2	100	0.2	24	365	8,760	1,752	4.87
Total					**6.8**		**4.1**				**35,916**	**100**

- Yearly energy consumption as the product of the consumed power multiplied by the working hours per year. If measurements are available, these figures can be used instead of estimates.
- Codes for identifying end users; generally four codes are enough to indicate user categories (process or facilities; production areas; product, center of cost, etc.) or their attribution to boiler plants, electrical substations, and other facility plants around the site.
- Share of the different production categories in the electric, thermal, and total energy consumption, with different degrees of detail depending on the available data and on their mode of elaboration (see models in Tables 17.5, 17.6, and 17.7).

4. *KPI evaluation.* KPI for the energy consumption of process and facility plants must be calculated on the basis of measurements. They can be used as a baseline for evaluating energy saving (see Table 17.8).

5. *Feasibility studies for investments.* Investment classification is shown in Table 17.9. Investment cost, savings (saving in energy and operating costs plus or minus additional costs) and a preliminary economic analysis by means of the pay-back parameter are shown in detail as well as the state of each investment evaluation (see also Chap. 19). Investments on centralized control of process and facility plants must be carefully investigated (see Sect. 17.9).

6. *Personnel training.* This is an important aspect of energy management and must be developed at different levels by including all the personnel (see also Chap. 18).

The models must be cross-checked to verify the consistency of different modes of water and energy consumption (electric, thermal, etc.) by the same end user. In this way, anomalies due to variation from standards can be detected.

The results of the model elaboration can also be used to break down energy consumption into the different production categories.

Energy flow measurements, which are made systematically in many parts of chemical, petrochemical, and other energy-intensive factories, can also be implemented in medium energy intensive manufacturing industries and in buildings by accurately choosing the main flows to be measured in order to reduce investments for instrumentation and related maintenance costs. Moreover, the results of these measurements can be used to improve the validity of the energy models (see step 3c) and the precision with which energy consumption is shared between different categories of production or services.

The modeling, which emphasizes the relative importance of different users and facilities from the energy point of view, is a preliminary phase for any energy accounting, whether off-line or on-line, and for any centralized energy management and plant control systems.

Table 17.8 Energy KPI performance indicators

	Unit	Site XXXX	Benchmark Minimum	Maximum	Installed instruments
Boiler plant					
Boiler plant efficiency (reference Higher Heating Value for US, Canada, ...)		80 %	80 %	85 %	
Boiler plant efficiency (reference Lower Heating Value)		90 %	90 %	95 %	
kg of steam referred to the unit of input natural gas	kg/Sm^3	11	10	12	
Cooling plants and cooling tower					
COP of cooling plants		4	3	5	
Water consumption referred to electrical consumption of the chillers	m^3/kWh	0.011	0.012	0.015	
Electric energy consumption of tower pumps and fans referred to the electric energy consumption of the chillers	%	14 %	5 %	10 %	
Compressed air plant					
Electrical consumption of the air compressors referred to inlet air	kWh/Sm^3	0.142	0.10	0.12	
HVAC—AHU					
Fan electrical consumption referred to building volumes	kWh/m^3	7.5	10	30	
Hourly cost of AHU (fan + cooling energy + heating energy)—reference AHU with a flow rate of 100,000 m^3/h	U/h	20	10	30	
Water					
Yearly sanitary water consumption referred to average manpower unit per day	m^3/unit	16	15	20	
Production					
Electric energy consumption referred to the production unit	kWh/t	1,000.7			
Natural gas consumption referred to the production unit	Sm^3/t	180.5			
Water consumption referred to the production unit	m^3/t	3.6			
Total energy cost referred to the production unit	U/t	148.8			

Table 17.9 Energy saving proposals—XXXX–YYYY

E:equipment M:management / P:process F:facilities (1)	Proposal description	Capital investment (A) Total (U)	Savings (B) Cost saving (U/year)	Electric energy (MWh/year)	Natural gas (Sm³/year)	Water (m³/anno)	O & M (U/year)	Primary energy (TOE/year) (2)	(C) Public aids (U/year)	(A)/(B) Payback (year)	IRR (%)
E	Cogeneration plant—1,000 kW output electric power and 1,000 kW hot water 60 °C/estimated saving 30 % of the current electric energy cost	1,000,000	310,340							3.22	
F											
E	Air handling units—better management of the control of temperature and humidity with a range of values/estimated saving 20 % of electrical and heat and cooling consumption	200,000	70,080							2.85	
F											
E	Compressed air—installation of 100 kW variable speed drive air compressor VFD in order to reduce no-load hours/estimated saving 15 % of the current electric energy consumption	120,000	34,161							3.51	
F											
Total		**1,320,000**	**414,584**	10,007.24	1,805,339	35,916	Energy consumption / Energy saving (%)			**3.18**	

Reference figures for technical and economic evaluations:

To convert from electric energy to primary energy multiple by	0.187	TOE/MWhe
To convert from Sm³ of natural gas primary energy multiple by	0.000825	TOE/Sm³
LHV of natural gas	34,450	kJ/Sm³
Electrical energy cost per unit	0.1	U/kWh
Natural gas cost per unit	0.25	U/Sm³
Water cost per unit	1	U/m³

Energy consumption U/year; Energy saving (%); Energy costs U/year; Energy cost savings (%) 27.86 %

Notes

(1) Process: the energy saving proposal refers to process equipment. Facilities: the energy saving proposal refers to facility equipment

(2) TOE (ton of oil equivalent); the figures refer to the conversion factors listed on the left of this table

(3) Public aids: public aids for energy saving investments

17.3 Preliminary Technical and Economic Evaluation of Energy-Saving Investments

The chief investments aimed at reducing energy costs and consumption must be submitted to a technical and economical analysis following the results of the previous steps. This is one of the main goals of an energy analysis and is often the more interesting phase for most people working in the site. Various methods of evaluation can be applied, for example those shown in Chap. 19. It is worth noticing that quite often energy-saving investments involve many other factors, such as reduction of maintenance costs and air pollution, the incidence of which is not easy to evaluate, but which often improves the return on the investment.

Investments can be classified as in Table 17.9, generally presented as the first part of an energy analysis report to stress the importance of these items.

> Note that investments on process plants must be checked carefully with the production technologists, because they can require important changes in the process, in both quality and quantity. Except for chemical, petrochemical, and other energy-intensive industries, it is difficult to justify investment on process plant only from the energy viewpoint; however, plant automation, improvement in production and replacement of obsolete plants are often the key factors in such decisions. On the other hand, it is easier to find an agreement on energy-saving investments on facilities in factories and buildings, since they do not affect process nor involve overall strategies directly. Therefore, the energy factor alone is often enough to account for investments on facilities, which may prove profitable and at the same time meet production needs.

17.4 Off-Line and On-Line Procedures for Energy Accounting

> The energy consumption trends are checked periodically by examining the energy accounting so as to verify data consistency and detect variance from standard values and company goals (for the whole process, for single lines, and for facility plants).

This procedure can be entrusted either to the company's internal structures or to independent organizations by means of on-line procedures (as in Sect. 17.8) or off-line procedures.

As a rule, these procedures include:
- Periodical elaboration (daily, weekly, monthly) of energy consumption and production data to update significant consumption parameters;

- Meetings with staff for critically evaluating consumption trends and checking the efficiency and development of company policies;
- Controller's reports providing an up-to-date picture of the energy situation and comments on observed anomalies. These reports can be diversified to suit the company functions they are intended for.

Energy accounting ensures a continuously high level of energy awareness and allows operating anomalies to be detected very quickly.

The depth and detail of accounting procedures depend on the size and complexity of the company's energy use and on the main motive determining the collection of data (basically, to break down only the energy cost by different categories of production or to achieve the greatest efficiency in energy use).

Data collection can be implemented off-line or on-line depending on the availability of centralized control for process and facilities plants. The achievement of better operation and maintenance is the main reason for having a centralized control equipment; in addition, it can conveniently be used for accounting procedures.

17.5   Feasibility Studies and Financing

Proposals for plant investments should start from a feasibility study with a technical and economical evaluation of the proposed solutions, also with a view to the company's present and future needs.

For project execution then it is necessary to choose the best financing channels and go through all the bureaucratic steps, taking into account energy and environmental regulations.

In many factories and buildings, a classic energy saving action focuses on heat recovery from water or hot gas streams or from cooling plant condensation. Investments in centralized plant control (see Sect. 17.8), particularly in facilities, require a preliminary study of specific energy parameters, measurement methods and control algorithms. For this purpose, the energy flows of the whole site as well as of single process lines must be known, and hence the need for combined strategies like those described in the previous paragraphs.

> **Non-energy additional costs and savings must be included in feasibility studies, particularly for the replacement of process and facility plants and for computer control systems. Very often, only the contribution of these savings can justify the investment.**

An economic evaluation must always be made. The principal methods are reviewed in Chap. 19.

17.6   Personnel Training

This is a further not be overlooked aspect of energy management, because energy-saving policies require responsive and motivated staff (see also Chap. 18).

Training is fundamental at all levels of company functions, plant operators, production planning and accounting operators, who often do not receive sufficient information or fail to see their own role and tasks in the context of an economical energy policy.

17.7   A Successful Energy Management Program

The success of an energy management program depends on many activities which must be promoted by the energy management staff and accepted and developed by all the personnel working on site.

A thorough knowledge of the whole site, in terms of process and facilities plants and of their operation and maintenance, is the first requirement. It is very useful making a list of all the energy saving opportunities, whether in plant operation or retrofitting or in new installations. The energy management staff should be aware of how production and environmental conservation are interconnected and should check the results of these investigations with other departments (production, maintenance, engineering, finance) before presenting a final report.

Advantages and disadvantages must be clearly indicated and an economic evaluation must always be included.

A periodical account of energy flows related to production must be organized before promoting any investment on plants. In this way, the results will be easy to check and the success of the energy strategies will be better appreciated.

Information on energy strategies and results must be made known throughout the site to motivate people to maintain the level of efficiency already reached and to improve it in the future.

17.8   Centralized Control of Process and Facility Plants

In order to be effective a centralized control requires a close analysis of process and facility plants. An energy audit is needed to supply the necessary information on the use of energy throughout the site and to identify the key operating parameters to be measured and controlled.

> **The use of computerized systems enhances process and facilities control mainly affecting four important functions: Operation, Maintenance, Planning, and Management. As energy comes into each of these functions, an overall approach dealing with every aspect of the site's life increases the opportunities for energy saving. In addition, the control system can be extended beyond the site's boundary so as to integrate end user customers, distributed generation from fuels and renewables, distribution and transmission networks, and utility generation plants. This global approach, usually called Smart Grid, enhances also the application of the best performing technologies.**

A detailed analysis of different control systems and their architecture is here omitted; specialized books and papers on information science and technical documentation from system manufacturers should be consulted to ensure that the chosen system is up-to-date and is best suited to the user's needs.

A basic scheme is shown in Fig. 17.2.

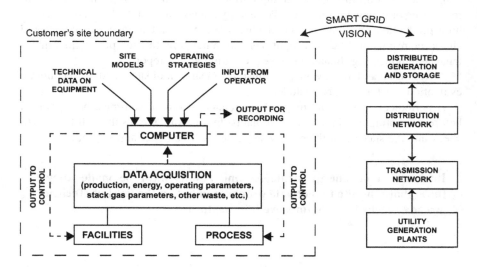

Fig. 17.2 A basic scheme for energy and waste management by computerized systems on site and interaction with external networks and utility plants

A centralized control system for energy saving, whatever its architecture and the available HW and SW, collects data from plants, processes, and facilities throughout the site. It analyzes and elaborates these data so as to activate operation strategies defined by the user and designed to maintain the best operating conditions with a minimum energy expenditure.

> **Sensors and transducers play an important role and very often represent the major part of the installation cost. To avoid installing a huge number of such devices it is particularly important choosing the essential parameters to measure and control through the energy audit.**

As a general rule, a standard SW program performs the following main functions:
- It acquires data from sensors by scanning individual frequencies of acquisition for various quantities (temperature, electric energy, pressure, steam flow, natural gas flow, water flow, exhaust flow, number of occupants, etc.). Subsequent processing converts the readings into engineering values (K or °C or °F, kW, t of steam, etc.). The maximum and minimum limits of the reading and the functioning of the instrumentation can usually be checked. The control system detects the on/off state of equipment such as valves, switches, etc. and also of the alarms. If required by the energy program, other quantities such as efficiency, enthalpy, coefficient of performance, etc., can periodically be calculated. The data collected and subsequently elaborated are generally memorized and preserved for 24 h or longer, according to the kind of operation.
- It chooses the operating modes on the basis of standard strategies already present or of strategies introduced by the users in particular situations, and it gives the output signals to activate the machines.

  Basically, strategies can be classified as operation strategies and optimization strategies. The first usually serve to bring an anomalous situation back to correct operational limits or to run an installation according to predetermined optimal criteria. The second, optimization strategies, are generally more complex and cause variables to interact with one another in such a way that the whole plant (total plant, individual production line, individual facility, etc.) is always working at maximum efficiency.
- It gives the operator periodical and on demand information on past data, current data, daily-monthly-yearly data, maintenance reports, and other reports for particular needs. Communication between operator and system must be as simple as possible in order to facilitate the full exploitation of the control system, which is generally under-used because of difficulties of interaction with operator.

The main facility-control strategies concern boiler plants, air compressor plants, pumps and fans, cooling plants, HVAC, and lighting as already detailed in the previous chapters. No indications can be given about process, because the great variety of cases would require specialized knowledge of multifarious sectors.

17.9   Practical Examples

An example of computer application to energy management in a site requires a great many details and data on process and facility equipment and on the hardware and software chosen.

There are many kinds of software and hardware available on the market, which make many levels of control possible.

Example   A computerized system to control a factory's facilities.

In order to give a general example of application in industry, in what follows the factory is represented by its electric and thermal model, as detailed in Table 17.10, which gives estimates of energy saving in facilities on the basis of practical experience.

Other advantages, such as reduction of working hours for plant maintenance and operation, will certainly increase the economic advantages of the control system (see also Example 2 in Sect. 13.7), but they are not here considered.

Economic evaluation is shown in Table 19.4.

Table 17.10   Energy saving due to computer control of facilities

	Average absorbed power kW	Working hours h/year	Annual consumption kWh/year	Estimated energy saving (%)	(kWh/year)
Electric model					
Air-compressor plant	250	5,000	1,250,000	7	87,500
Pump plant	75	5,000	375,000	15	56,250
HVAC refrigerating plant	500	1,750	875,000	7.5	65,625
HVAC auxiliaries	100	3,500	350,000	12.5	43,750
Lighting	100	5,000	500,000	5	25,000
Process	800	5,000	4,000,000	0	0
Others	250	5,000	1,250,000	0	0
Total	*2,075*		*8,600,000*		*278,125*
	a	b	c = a × b	d	e = c × d/100
Thermal model				(%)	(TOE/year)
Process boiler plant	7,500	5,000	3,225	3	96.751
HVAC boiler plant	1,000	1,750	150.5	3	4.515
Total	*8,500*		*3,375.5*		*101.266*
	a	b	c (*)	d	e = c × d/100

Notes (*) c = a × b × 3,600/(41,860 × 10³) where LHV of oil is 41,860 kJ/kg

Education in Energy Conversion and Management

18

18.1 Introduction

Awareness of the role of energy in all aspects of life—home, industry, transport, and services—is a primary step in energy management education.

This concept can easily be introduced at any level of the educational system, but an appropriate presentation must be devised for each case. The energy and mass balance, which underlies the basic principle of energy conversion and management and allows the solution of many practical problems, can be explained in simple terms even in primary schools and can be a useful guideline for the comprehension of most phenomena.

Of course, the same principle will be better understood in the following years, when physics, chemistry, mathematics, thermodynamics, and other specific courses will be encountered and the different aspects of energy transformation will be clarified also by practice at work.

General education must highlight the broad aspects of energy transformation; technical schools and universities must provide a detailed analysis of components and systems and finally a general overview to connect all these elements.

18.2 The Role of University

According to the tradition of each country, university is oriented to give either a specialized or a more broadly based formation. In both cases, all the systems in which energy transformation occurs are the subject of courses, such as electrical and thermal sciences, thermodynamics including heat transfer and fluid mechanics, and the subsequent specialized courses on machines and plants. Other courses, not specifically related to energy, such as power electronics, information science, and industrial automation, are generally offered to complete the student's curriculum. The university has to coordinate these courses to give students an exhaustive view of the various concepts and prepare them for their future work.

> University is the right place to present the principles of energy transformation free from constraints due to production requirements and economic evaluations. The economic problems are not overlooked, but priority is given to technical aspects.

A clear and complete understanding of the technical facets of a problem is the basis for assessing all the other factors, economic and human, which will make up the decision criteria in industry, but which cannot be treated without regard for technical considerations.

Energy and mass balance can be assumed as the keystone for a wide comprehension of any phenomenon; if thoroughly understood, this concept becomes the basic means to solve any problem one may have to face in energy management.

18.3 Personnel Training

> Continuous training and motivation of both graduates and other technicians is the main key to the success of any energy management project.

Training means giving people of all ranks energy-related information to assist them in their day-to-day work and planning.

Motivation means pushing people by economic and human incentives to feel very concerned about energy saving as a high priority.

Courses on different energy-management topics can be organized at various levels, depending on the background and staff position of each employee. However, as a primary goal all workers must know what they are working for, how their contribution is used, and what the main results of the project are.

Meetings for the presentation and discussion of energy-management projects and results of energy-accounting procedures can also constitute part of the personnel training program.

18.4 Awareness of Energy Conversion and Management as an Intersectoral Discipline

Figure 18.1 shows a matrix presentation of the interaction between energy conversion and management topics and the specialized courses traditionally offered in many universities and technical schools. The complexity of the interaction is evident.

Although energy-related topics have often been dealt with as parts of different cultural areas, energy management involves so many technical and economic aspect of life that it is now worth approaching it as an independent discipline.

The integration of several topics to learn in detail, from basic concepts to plant design parameters, highlights the need for a systematically arranged program of studies which includes many traditional courses, but pays particular attention to energy-transformation concepts. This means changing the approach to the topics of each individual course by emphasizing this common basis and by preparing students to focus on it. If necessary, individual courses can be integrated with one another by reorganizing the program so as to offer a specific energy conversion and management curriculum.

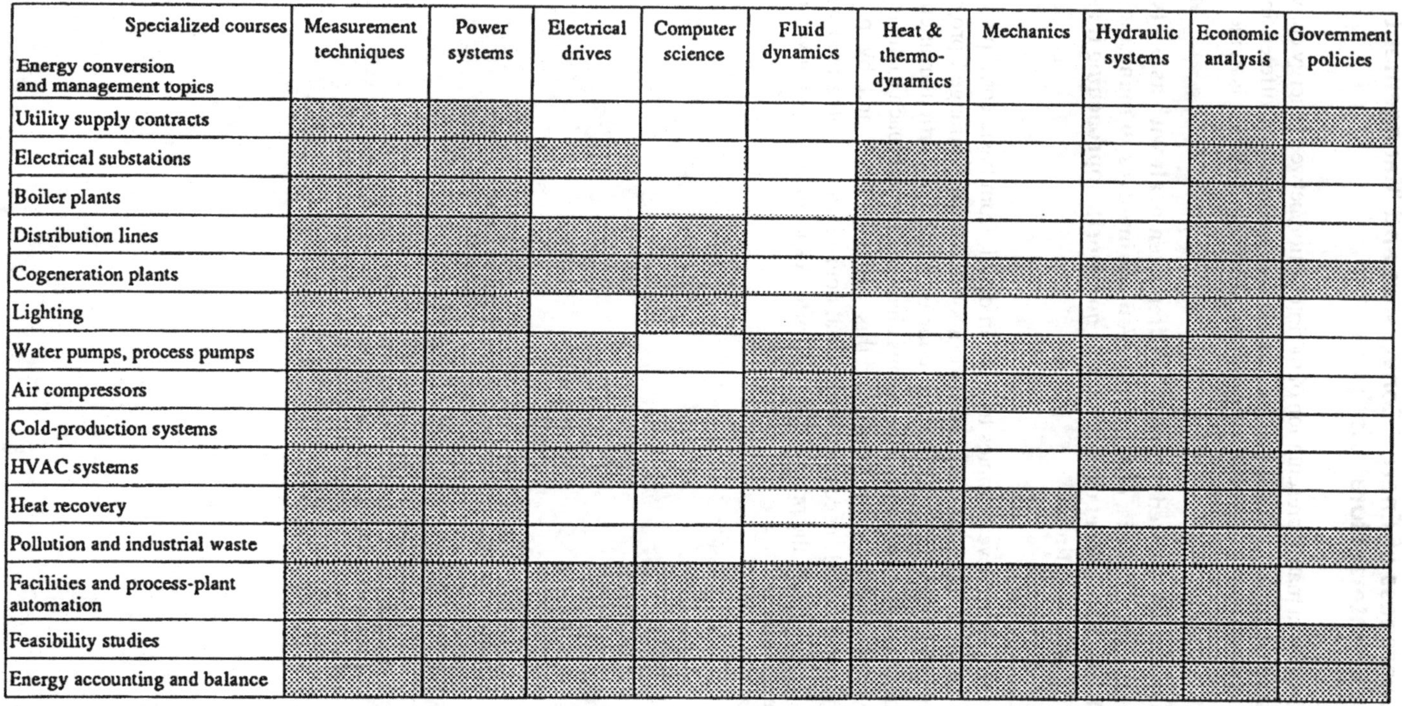

Specialized courses / Energy conversion and management topics	Measurement techniques	Power systems	Electrical drives	Computer science	Fluid dynamics	Heat & thermo-dynamics	Mechanics	Hydraulic systems	Economic analysis	Government policies
Utility supply contracts	■	■							■	■
Electrical substations	■	■	■			■			■	
Boiler plants	■					■			■	
Distribution lines	■			■				■	■	
Cogeneration plants	■			■			■	■	■	■
Lighting	■								■	■
Water pumps, process pumps	■		■	■	■		■	■	■	■
Air compressors	■		■	■			■	■	■	
Cold-production systems	■		■	■	■	■	■	■	■	
HVAC systems	■			■		■	■	■	■	
Heat recovery	■					■	■	■	■	
Pollution and industrial waste	■					■	■	■	■	■
Facilities and process-plant automation	■		■	■	■	■	■	■	■	
Feasibility studies	■			■					■	■
Energy accounting and balance	■			■					■	■

Fig. 18.1 Integration between energy conversion and management topics and specialized courses (*shaded areas* evidence integration)

Economic Analysis of Energy-Saving Investments

19

19.1 Introduction

The economic analysis of investments is a critical step in an energy conservation program because monetary saving is generally the main factor leading to a decision. The same analysis can be conveniently used to choose among possibilities which may be equivalent from a technical point of view.

The main elements of an investment are the capital cost or the initial investment, the interest rate, the return on the investment, and the life of the investment.

There are several methods available, according to the company's internal evaluation criteria for investment, not only in the energy-saving field. Depending on the importance and on the life of the investment, the selected methods can include or not include life cycle costing as well as more or less sophisticated approaches.

A short review of the main analytical methods applicable to energy-saving investments is reported as follows.

19.2 Methods Not Using Life Cycle Costing

19.2.1 Payback Method

> The payback period is the time required to recover the capital investment from net cash flow.

Payback = INVESTMENT/NET ANNUAL CASH FLOW = I/NACF

where investment I = the total capital cost, and net annual cash flow NACF = (annual energy cost saving + other cost saving − annual additional costs) referred to the year of the investment.

G. Petrecca, *Energy Conversion and Management: Principles and Applications*, DOI 10.1007/978-3-319-06560-1_19, © Springer International Publishing Switzerland 2014

The two terms of the payback ratio can be either before or after taxes depending on the requirements of the investor.

If the salvage cost, that is, the residual value of the equipment at the end of its useful life, is deducted from the initial capital cost, a payback with salvage is calculated. Of course, it is shorter than the payback without salvage as defined above.

The payback method does not consider the savings after the payback years; thus it penalizes projects that have a long potential life in comparison with those that offer high savings for a short period.

The payback method does not consider energy-pricing variation or the time value of the money.

Nevertheless, the payback method is very simple and it can serve as a yardstick to compare possible investments.

19.2.2 Investors' Rate of Return (ROR) Method

$$\text{ROR } (\%) = \text{NET ANNUAL CASH FLOW/INVESTMENT} = \text{NACF}/I.$$

This is the reciprocal of the payback and it is generally expressed as a percentage.

19.3   Methods Using Life Cycle Costing

These methods are based on the conversion of investment and annual cash flow at various times to their equivalent present values and vice versa. The real interest or discount rate (r), that is, the nominal discount rate less the inflation rate, and the number of years (n) of the evaluation period have to be considered.

Several factors are used to accomplish these conversions:

- Future worth factor:

     This converts a single present amount (at year zero) to an amount at a future point in time:

$$\text{FWF} = (1 + r)^n.$$

- Present worth factor:

     This converts a future amount to an amount today (at year zero):

$$\text{PWF} = \frac{1}{(1 + r)^n}$$

19.3.1 Present Worth Method

> **The net present worth of a project is defined as the difference between the present worth of the total project revenues (energy cost saving + other cost savings – additional operating costs) and the present capital cost of the project:**

$$PW = \sum_{1}^{n} j \frac{1}{(1+r)^j} \left(C_{pel} \times Q_{el_j} + C_{P_{th}} \times Q_{th_j} \pm C_{pa} Q_{a_j} \right) - I_p$$

where C_p = present monetary value (at year zero) of each unit of revenue (C_{pel}: value of unit of electrical energy saving; $C_{P_{th}}$: value of unit of thermal energy saving; C_{pa}: value of unit of additional saving or cost of materials and working hours), and Q_j = annual revenues in physical units (Q_{el_j}: amount of electric energy saving; Q_{th_j}: amount of thermal energy saving; Q_{a_j}: amount of additional or saved materials and working hours, etc.). It is usually assumed that they become effective in year 1 ($j = 1$); I_p = present investment at year zero. If the investment has been made in different years, the present investment can be calculated by using the PWF factor:

$$= I_o + \sum_{1}^{n} j \frac{I_j}{(1+r)^j}$$

In the case of energy-saving investments, all savings and additional operating costs can also be expressed as a percentage of the energy saving by increasing or decreasing the present monetary value of the energy (C_{pel}, $C_{P_{th}}$).

The present investment I_p can be simply the effective capital cost if it is concentrated in the year zero or the present worth of the investment calculated by means of the present worth factor if it is made in several years.

The expression given above can be adapted for wider use if a different inflation rate is introduced for each item of the revenues and for the capital investment. A widely accepted criterion takes into account two different values of the inflation rate for the energy and related revenues (f^*) and for the investment (f).

A general expression for PW is

$$
\begin{aligned}
\text{PW} &= \sum_{1}^{n} j \frac{\left(C_{\text{pel}} \times Q_{\text{el}_j} + C_{\text{p}_{\text{th}}} Q_{\text{th}_j} \pm C_{\text{pa}} Q_{\text{a}_j}\right) \times (1+f^*)^j}{(1+r)^j \times (1+f)^j} - I_{\text{p}} \\
&= \sum_{1}^{n} j \frac{\left(C_{\text{pel}} \times Q_{\text{el}_j} + C_{\text{p}_{\text{th}}} Q_{\text{th}_j} \pm C_{\text{pa}} Q_{\text{a}_j}\right)}{(1+i)^j} - I_{\text{p}}
\end{aligned}
$$

where $i = r + f - f^*$ for small values of r, f, f^*.

In order to obtain a simpler expression, annual energy saving in physical units is assumed constant during the life of the investment, and so are other savings and additional expenditures. In addition, the present investment I_{p} is assumed to be equal to the total investment I, which is taken to occur wholly in the present year (year zero).

The simplified PW expression becomes as follows:

$$
\begin{aligned}
\text{PW} &= \left(C_{\text{pel}} \times Q_{\text{el}} + C_{\text{p}_{\text{th}}} Q_{\text{th}} \pm C_{\text{pa}} Q_{\text{a}}\right) \times \sum_{1}^{n} j \frac{1}{(1+i)^j} - I \\
&= \left(C_{\text{pel}} \times Q_{\text{el}} + C_{\text{pth}} \times Q_{\text{th}} \pm C_{\text{pa}} Q_{\text{a}}\right) \times \text{PAF} - I = \text{NACF} \times \text{PAF} - I
\end{aligned}
$$

where Q_{el}, Q_{th}, and $Q_{\text{a}} =$ annual amount of revenues in physical units assumed constant throughout the life of the investment. They start to become effective in year 1; $n =$ years of investment life, $i = r + f - f^*$ ($i = r$, if $f = f^*$), and $I =$ total investment concentrated in the year zero (see Payback):

$$
\sum_{1}^{n} j \frac{1}{(1+i)^j} = \text{Present annuity factor (PAF)}
$$

NACF = Net annual cash flow (see Payback).

If PW is greater than zero, the project is valid since the revenues are enough to pay the interest and to recover the initial capital cost before the end of the life of the investment. If PW equals zero, the balance occurs at the end of the life, but the investment is scarcely attractive. PW less than zero means that the project is a bad one.

Projects can conveniently be compared by taking as a parameter the ratio between the present worth of the project and the related investment (PW/I).

Factors such as PWF and PFA can be evaluated either by using tables (see Tables 19.1 and 19.2) or by calculation on personal computers, with any software available.

Table 19.3 shows an example of PW calculation by using both the general and the simplified expressions.

19.3.2 Internal Rate of Return (IRR)

IRR is the value of i) such that $PW = 0 = NACF \times PAF - I$.

It follows that $PW = 0$ if $PAF = NACF/I = PAYBACK$.

> **IRR is the internal interest rate that reduces to zero the present worth of the project at the end of the life of the project (n years).**

This parameter, named IRR, can easily be compared with the company interest rate to evaluate the merit of the project. The calculation of IRR requires an iterative approach: selecting different values of the parameter i) and calculating the PW, generally using the basic expression.

19.4 Case Studies

Several examples discussed from the technical point of view in the previous chapters are here examined to evaluate the economic validity of the correlated investment from the cost energy-saving angle.

The basic methods reported in Sects. 19.2 and 19.3 are used and the results for all the case studies examined throughout the book are shown in Table 19.4.

To facilitate the understanding the table provides example values for the energy costs and for the investment. Notice that the ratio between electric and thermal energy costs, which is introduced as an example, has a general validity because this ratio is roughly the same for many countries. The cost of unit of energy (MWh for electric energy, TOE for thermal energy) is expressed as kU/MWh or kU/TOE where U can be any currency; a reference cost of 0.1 kU/MWh and 0.16 kU/TOE of thermal energy is assumed.

For each case, Table 19.4 shows the annual energy saving, both electrical and thermal, and the corresponding cost saving. Expected investment life is also introduced; investment I is expressed in kU units.

The different kinds of investment are compared by using payback and IRR parameters.

Notice that only energy cost and saving have been considered; for a more detailed analysis, additional costs and savings (working hours, maintenance,

Table 19.1 Present worth factor values PWF $= \frac{1}{(1+i)^n}$

Year (1)	PWF values corresponding to coefficient i in %									
	1 %	2 %	3 %	4 %	5 %	6 %	7 %	8 %	9 %	10 %
1	0.990	0.980	0.971	0.962	0.952	0.943	0.935	0.926	0.917	0.909
2	0.980	0.961	0.943	0.925	0.907	0.890	0.873	0.857	0.842	0.826
3	0.971	0.942	0.915	0.889	0.864	0.840	0.816	0.794	0.772	0.751
4	0.961	0.924	0.888	0.855	0.823	0.792	0.763	0.735	0.708	0.683
5	0.951	0.906	0.863	0.822	0.784	0.747	0.713	0.681	0.650	0.621
6	0.942	0.888	0.837	0.790	0.746	0.705	0.666	0.630	4.596	0.564
7	0.933	0.871	0.813	0.760	0.711	0.665	0.623	0.583	0.547	0.513
8	0.923	0.853	0.789	0.731	0.677	0.627	0.582	0.540	0.502	0.467
9	0.914	0.837	0.766	0.703	0.645	0.592	0.544	0.500	0.460	0.424
10	0.905	0.820	0.744	0.676	0.614	0.558	0.508	0.463	0.422	0.386
11	0.896	0.804	0.722	0.650	0.585	0.527	0.475	0.429	0.388	0.350
12	0.887	0.788	0.701	0.625	0.557	0.497	0.444	0.397	0.356	0.319
13	0.879	0.773	0.681	0.601	0.530	0.469	0.415	0.368	0.326	0.290
14	0.870	0.758	0.661	0.577	0.505	0.442	0.388	0.340	0.299	0.263
15	0.861	0.743	0.642	0.555	0.481	0.417	0.362	0.3 15	0.275	0.239
16	0.853	0.728	0.623	0.534	0.458	0.394	0.339	0.292	0.252	0.218
17	0.844	0.714	0.605	0.513	0.436	0.371	0.317	0.270	0.231	0.198
18	0.836	0.700	0.587	0.494	0.416	0.350	0.296	0.250	0.212	0.180
19	0.828	0.686	0.570	0.475	0.396	0.331	0.277	0.232	0.194	0.164
20	0.820	0.673	0.554	0.456	0.377	0.312	0.258	0.215	0.178	0.149
	11 %	12 %	13 %	14 %	15 %	16 %	17 %	18 %	19 %	20 %
1	0.901	0.893	0.885	0.877	0.870	0.862	0.855	0.847	0.840	0.833
2	0.812	0.797	0.783	0.769	0.756	0.743	0.731	0.718	0.706	0.694
3	0.731	0.712	0.693	0.675	0.658	0.641	0.624	0.609	0.593	0.579
4	0.659	0.636	0.613	0.592	0.572	0.552	0.534	0.516	0.499	0.482
5	0.593	0.567	0.543	0.519	0.497	0.476	0.456	0.437	0.419	0.402
6	0.535	0.507	0.480	0.456	0.432	0.410	0.390	0.370	0.352	0.335
7	0.482	0.452	0.425	0.400	0.376	0.354	0.333	0.3 14	0.296	0.279
8	0.434	0.404	0.376	0.351	0.327	0.305	0.285	0.266	0.249	0.233
9	0.391	0.361	0.333	0.308	0.284	0.263	0.243	0.225	0.209	0.194
10	0.352	0.322	0.295	0.270	0.247	0.227	0.208	0.191	0.176	0.162
11	0.317	0.287	0.261	0.237	0.215	0.195	0.178	0.162	0.148	0.135
12	0.286	0.257	0.231	0.208	0.187	0.168	0.152	0.137	0.124	0.112
13	0.258	0.229	0.204	0.182	0.163	0.145	0.130	0.116	0.104	0.093
14	0.232	0.205	0.181	0.160	0.141	0.125	0.111	0.099	0.088	0.078
15	0.209	0.183	0.160	0.140	0.123	0.108	0.095	0.084	0.074	0.065
16	0.188	0.163	0.141	0.123	0.107	0.093	0.081	0.071	0.062	0.054
17	0.170	0.146	0.125	0.108	0.093	0.080	0.064	0.060	0.052	0.045
18	0.153	0.130	0.111	0.095	0.081	0.069	0.059	0.051	0.044	0.038
19	0.138	0.116	0.098	0.083	0.070	0.060	0.051	0.043	0.037	0.031
20	0.124	0.104	0.087	0.073	0.061	0.051	0.043	0.037	0.031	0.026

Table 19.2 Present annuity factor values $\mathsf{PAF} = \sum_{1}^{n} j \frac{1}{(1+i)^j}$

Year (j)	PAF values corresponding to coefficient i in %									
	1 %	2 %	3 %	4 %	5 %	6 %	7 %	8 %	9 %	10 %
1	0.990	0.980	0.971	0.962	0.952	0.943	0.935	0.926	0.917	0.909
2	1.970	1.942	1.913	1.886	1.859	1.833	1.808	1.783	1.759	1.736
3	2.941	2.884	2.829	2.775	2.723	2.673	2.624	2.577	2.531	2.487
4	3.902	3.808	3.717	3.630	3.546	3.465	3.387	3.312	3.240	3.170
5	4.853	4.713	4.580	4.452	4.329	4.212	4.100	3.993	3.890	3.791
6	5.795	5.601	5.417	5.242	5.076	4.917	4.767	4.623	4.486	4.355
7	6.728	6.472	6.230	6.002	5.786	5.582	5.389	5.206	5.033	4.868
8	7.652	7.325	7.020	6.733	6.463	6.210	5.971	5.747	5.535	5.335
9	8.566	8.162	7.786	7.435	7.108	6.802	6.515	6.247	5.995	5.755
10	9.471	8.983	8.530	8.111	7.722	7.360	7.024	6.710	6.418	6.145
11	10.368	9.787	9.253	8.760	8.306	7.887	7.499	7.139	6.805	6.495
12	11.255	10.575	9.954	9.385	8.863	8.384	7.943	7.536	7.161	6.814
13	12.134	11.348	10.635	9.986	9.394	8.853	8.358	7.904	7.487	7.103
14	13.004	12.106	11.296	10.563	9.899	9.295	8.745	8.244	7.786	7.367
15	13.865	12.849	11.938	11.118	10.380	9.712	9.108	8.559	8.061	7.606
16	14.718	13.578	12.561	11.652	10.838	10.106	9.447	8.851	8.313	7.824
17	15.562	14.292	13.166	12.166	11.274	10.477	9.763	9.122	8.544	8.022
18	16.398	14.992	13.754	12.659	11.690	10.828	10.059	9.372	8.756	8.201
19	17.226	15.678	14.324	13.134	12.085	11.158	10.336	9.604	8.950	8.365
20	18.046	16.351	14.877	13.590	12.462	11.470	10.594	9.818	9.129	8.514
	11 %	12 %	13 %	14 %	15 %	16 %	17 %	18 %	19 %	20 %
1	0.901	0.893	0.885	0.877	0.870	0.862	0.855	0.847	0.840	0.833
2	1.713	1.690	1.668	1.647	1.626	1.605	1.585	1.566	1.547	1.528
3	2.444	2.402	2.361	2.322	2.283	2.246	2.210	2.174	2.140	2.106
4	3.102	3.037	2.974	2.914	2.855	2.798	2.743	2.690	2.639	2.589
5	3.696	3.605	3.517	3.433	3.352	3.274	3.199	3.127	3.058	2.991
6	4.231	4.111	3.998	3.889	3.784	3.685	3.589	3.498	3.410	3.326
7	4.712	4.564	4.423	4.288	4.160	4.039	3.922	3.812	3.706	3.605
8	5.146	4.968	4.799	4.639	4.487	4.344	4.207	4.078	3.954	3.837
9	5.537	5.328	5.132	4.946	4.772	4.607	4.451	4.303	4.163	4.031
10	5.889	5.650	5.426	5.216	5.019	4.833	4.659	4.494	4.339	4.192
11	6.207	5.938	5.687	5.453	5.234	5.029	4.836	4.656	4.486	4.327
12	6.492	6.194	5.918	5.660	5.421	5.197	4.988	4.793	4.611	4.439
13	6.750	6.424	6.122	5.842	5.583	5.342	5.118	4.910	4.715	4.533
14	6.982	6.628	6.302	6.002	5.724	5.468	5.229	5.008	4.802	4.611
15	7.191	6.811	6.462	6.142	5.847	5.575	5.324	5.092	4.876	4.675
16	7.379	6.974	6.604	6.265	5.954	5.668	5.405	5.162	4.938	4.730
17	7.549	7.120	6.729	6.373	6.047	5.749	5.475	5.222	4.990	4.775
18	7.702	7.250	6.840	6.467	6.128	5.818	5.534	5.273	5.033	4.812
19	7.839	7.366	6.938	6.550	6.198	5.877	5.584	5.316	5.070	4.843
20	7.963	7.469	7.025	6.623	6.259	5.929	5.628	5.353	5.101	4.870

Table 19.3 Comparison between general and simplified PW relationships

PW calculation by the general relationship

Year	Investment	Annual revenues	Yearly cash flow	PWF ($i = 10\ \%$)	Present cash flow	PW of the project
	kU	kU	kU		kU	kU
0	1,500	0	−1,500		−1,500.0	−1,500.0
1	500	400	−100	0.909	−90.9	−1,590.9
2	250	700	450	0.826	371.7	−1,219.2
3	0	700	700	0.751	525.7	−693.5
4	0	700	700	0.683	478.1	−215.4
5	0	700	700	0.621	434.7	219.3
At the end of the life						219.3
a	b	c	$d = c - b$	e	$f = d \times e$	$g = $ sum of yearly f

In the following PW calculation by the simplified relationship is reported

Average annual revenues in 5 years equal to $(400 + 700 + 700 + 700 + 700)/5 = 640$ kU

Total investment concentrated in the year zero equal to 2,250 kU

PAF (5 years, from year 1 to year 5; $i = 10\ \%$) equal to 3.791

PW $= 640 \times 3.791 - 2,250 = 176.2$ kU

Notes

Annual revenues include all kinds of savings and additional operating costs

Annual revenues become effective in the year 1

replacement of components during the investment life, etc.), which often markedly affect the economic evaluation, must be taken into account.

The same approach can be followed for any other energy-saving investment in process and facility plants, as it is quite a good way to correlate and to compare different investments inside a factory.

Table 19.4 Examples of case studies

Description of case studies	Section	Annual energy saving		Annual cost saving		Investment life	Investment	Economic evaluation parameters	
		Q_{el}	Q_{th}	$C_{pel} \times Q_{el}$	$C_{pth} \times Q_{th}$			PAYBACK	IRR
		MWh/year	TOE/year	kU/year		Year	kU	Year	%
High-efficiency transformer	5.4	4.6	0.0	0.5	0.0	12	5[a]	10.95	1.40
High-efficiency boiler	6.12	0.0	165.21	0.0	26.4	10	200[a]	7.56	4.5
Power factor control	7.7								
Table 7.2, Case 4		35.2	0.0	3.5	0.0	10	15	4.26	42.70
Table 7.3, Case 4		8.8	0.0	0.9	0.0	10	5	5.69	26.50
Microwave instead of thermal dryer	7.7	−48	113	−4.8	18.1	10	125[a]	11.3	1
High-efficiency electric motor	7.7	11.4	0.0	1.1	0.0	10	3	2.63	36.30
Wall insulation	8.5	0.0	23.0	0.0	3.7	10	4	1.09	92.00
Pipeline insulation	8.5	0.0	154.5	0.0	24.7	10	8	0.32	309.00
Cogeneration[b]									
Steam turbine	9.7	6,000.0	−705		428.1	10	1,200	2.8	34
Gas turbine	9.7	6,000.0	−764		387.7	10	1,300	3.35	27
Reciprocating engine	9.7	6,000.0	−628		409.6	10	1,000	2.4	40
Trigeneration[b]									
Reciprocating engine	9.7	7,176.0	−1,229		431.2	10	1,100	2.8	34
Non-dissipative pump regulation	10.6	20.5	0.0	2.0	0.0	10	5	2.44	39.50
Air compressor plant	11.7								
Pressure reduction		95.5	0.0	9.6	0.0	10	25	2.62	36.50
Heat recovery		0.0	18.1	0.0	2.9	10	10	3.45	26.10

(continued)

Table 19.4 (continued)

Description of case studies	Section	Annual energy saving Q_{el} MWh/year	Q_{th} TOE/year	Annual cost saving $C_{pel} \times Q_{el}$ kU/year	$C_{pth} \times Q_{th}$ kU/year	Investment life Year	Investment kU	Economic evaluation parameters PAYBACK Year	IRR %
Refrigeration plant	12.10								
Replacement of absorption system		−1,140	902.0	−114.0	144.3	10	305	10.06	–
Heat recovery		0.0	80.7	0.0	12.9	10	20	1.55	64.00
HVAC	13.7								
Recovery from exhaust air		−21.7	422	−2.2	67.5	10	600	9.2	1.5
Computer control		240.0	90.0	24.0	14.4	8	200	5.2	11
Lighting	14.6								
Replacement of fluorescent with HF		249.3	0.0	24.9	0.0	8	160	6.4	5.1
Replacement of HF with LED		135.8	0.0	13.6	0.0	12	100	7.35	8
Heat recovery	15.7								
Water/water		0.0	474.3	0.0	75.9	10	75	1.0	101
Air/air		0.0	84.0	0.0	13.4	5	25	1.9	45.5
Plant control	17.9	278.1	101.3	27.8	16.2	8	250	5.7	8

Assumptions: $C_{pel} = 0.1$ kU/MWh; $C_{pth} = 0.16$ kU/TOE; additional saving and costs are ignored

[a] Extra cost

[b] O&M cost considered

Conclusions

20

Having reached the end of the book, readers should now be well acquainted with energy transformations in every site, factory or building, and with the main possibilities of energy saving. Basic principles, of which vague memories often remain from wider studies in the past, should now be clear in their essential meaning and provide useful guidance for the comprehension of most phenomena. Presented throughout the book as the key to solve any problem, energy and mass balance should now be quite a familiar tool and the flow of energy, from site boundaries to end users, should be clearly understood in all its transformations.

> **Basic formulas, data, and KPIs that are the essentials of energy conversion and management—which all those involved in these fields should perfectly know—are reported in Tables 20.1, 20.2, and 20.3. Any problem can be understood and solved by reducing it to a formula in which only basic data is required. KPIs, which represent an updated picture of industrial and end user performances, are the tools for checking the results and comparing different situations.**

Students will have had an opportunity to connect various subjects scattered through the university curriculum.

Technicians, not specifically concerned in designing plants, will have had access to basic data for a quick evaluation of many problems.

Managers will have found a guide to the understanding of many investment proposals and maybe a spur to solicit new ones.

The author hopes to have achieved the aim of this book and that readers will have obtained both answers to their problems and a global comprehension of the energy flow in factories and buildings, or at least a clear recapitulation of energy conversion and management principles and a stimulus to study particular topics in more detail.

G. Petrecca, *Energy Conversion and Management: Principles and Applications*,
DOI 10.1007/978-3-319-06560-1_20, © Springer International Publishing Switzerland 2014

Table 20.1 Basic formulas

SI system				English system			
Heating							
c [kJ/kg × K]	× m [kg/s]	× ΔT [K]	[kW]	c [Btu/lb × °F]	× m [lb/h]	× ΔT [°F]	[Btu/h]
Δh [kJ/kg]	× m [kg/s]		[kW]	Δh [Btu/lb]	× m [lb/h]		[Btu/h]
Heat transfer							
A [m²]	× U [W/m² × K]	× ΔT [K]	[W]	A [ft²]	× U [Btu/h × ft² × °F]	× ΔT [°F]	[Btu/h]
A [m²]	× $(1/R_{\text{th}})$ [W/m² × K]	× ΔT [K]	[W]	A [ft²]	× $(1/R_{\text{th}})$ [Btu/h × ft² × °F]	× ΔT [°F]	[Btu/h]
Ideal gas							
p [Pa]	× v [m³]	= constant × T [K]		p [psi]	× v [ft³]	= constant × T [R]	
Water pump							
Q [m³/s]	× H [m]	× (9.81)/η	[kW]	Q [gpm]	× H [ft]	× (0.188)/η/1,000	[kW]
Hydraulic turbine							
Q [m³/s]	× H [m]	× (9.81) × η	[kW]	Q [gpm]	× H [ft]	× (0.188) × η/1,000	[kW]
Gas compressor (isentropic) (from isentropic to real compression multiplied by 1.5)							
q_{inlet} [m³/s]	× p_{inlet} [Pa]	× $\ln(p_{\text{outlet}}/p_{\text{inlet}})$/1,000	[kW]	q_{inlet} [ft³/min]	× p_{inlet} [psi]	× $\ln(p_{\text{outlet}}/p_{\text{inlet}})$/1,000	[kW]
						× (3.254)	
Fan							
Q [m³/s]	× ΔP/η/1,000 [Pa]		[kW]	Q [ft³/min]	× ΔP [in H$_2$0]	× (0.1175)/η/1,000	[kW]

Table 20.2 Basic data

	SI unit		English unit	
Specific heat-water	4.186	kJ/kg × K	1	Btu/lb × °F
Specific heat-air	1	kJ/kg × K	0.239	Btu/lb × °F
Specific heat-superheated industrial steam (0.5–1 MPa)	2.09	kJ/kg × K	0.499	Btu/lb × °F
Specific heat-steam in the atmospheric air (<0.1 MPa)	1.8	kJ/kg × K	0.430	Btu/lb × °F
Average enthalpy of industrial steam	2,600	kJ/kg	1,160	Btu/lb
Density − air[a]	1.29	kg/Sm^3	0.081	lb/Sft^3
Density − natural gas[b]	0.75	kg/Sm^3	0.047	lb/Sft^3
Density − water	1,000	kg/m^3	62.430	lb/ft^3
			8.35	lb/gallon
Lower Heating Value (LHV) − oil	41,860	kJ/kg	17,997	Btu/lb
Lower Heating Value (LHV) − natural gas[b]	34,325	kJ/Sm^3	922	Btu/Sft^3

[a]Referred to 0 °C, 32 °F and 0.1 MPa, 14.69 psi
[b]Referred to 15.6 °C, 60 °F and 0.1 MPa, 14.69 psi

Table 20.3 Basic KPIs

	By	SI unit		English unit	
Steam production	Boiler	12–14	kg_{steam}/kg_{oil}	12–14	lb_{steam}/lb_{oil}
Steam production	Boiler	9–11	$kg_{steam}/$ $Sm^3_{natural\ gas}$	0.25– 0.31	$lb_{steam}/$ $Sft^3_{natural\ gas}$
COP cooling above ice point	Cooling compressor	3–6		3–6	
COP cooling below ice point	Cooling compressor	0.7–2		0.7–2	
COP cooling above ice point	Cooling absorption	0.7–1		0.7–1	
Compressed air	Air compressor	0.11– 0.14	$kWh/$ $Sm^3_{inlet\ air}$	0.0031– 0.004	$kWh/Sft^3_{inlet\ air}$
COP heating	Heat pump-compressor	3–6		3–6	
Condensing water	Cooling tower	0.01– 0.02	$m^3/$ $kWh_{compressor}$	2.6–5.2	$gallon/$ $kWh_{compressor}$
Condensing water	Discharged water	0.1–0.2	$m^3/$ $kWh_{compressor}$	26–52	$gallon/$ $kWh_{compressor}$
Auxiliaries in cooling tower or air condenser	Fans and pumps	0.05–0.1	$kWh_{aux}/$ $kWh_{compressor}$	0.05–0.1	$kWh_{aux}/$ $kWh_{compressor}$
Utility plant	Condensing turbine plant	4.5	kWh/kg_{oil}	2	kWh/lb_{oil}
Utility plant	Combined gas turbine plant	5.3	$kWh/$ $Sm^3_{natural\ gas}$	0.15	$kWh/$ $Sft^3_{natural\ gas}$
Water storage	Pump	250–300	$t \times m/kWh$	216,680– 260,000	$gallon \times ft/kWh$
Lighting	Luminaire	50–100	lm/W	50–100	lm/W
Lighting a building	Luminaire	10–20	W/m^2	0.93– 1.86	W/ft^2
Lighting a factory	Luminaire	5–10	W/m^2	0.46– 0.93	W/ft^2
Heating and cooling a building	Heating and cooling	10–20	W/m^3	1,205– 2,410	$Btu/h \times ft^3$
Heating and cooling a factory	Heating and cooling	5–10	W/m^3	602– 1,204	$Btu/h \times ft^3$

Volume of natural gas usually referred to 15.6 °C, 60 °F and 0.1 MPa, 14.69 psi

Volume of inlet air for compressed air plant referred to 15.6 °C, 60 °F and 0.1 MPa, 14.69 psi

References

Various Authors (2004) Encyclopedia of energy. Elsevier Science Publishing
Albert Thumann, P.E., C.E.M (1987) Plant engineers and managers guide to energy conservation, 3rd edn. USA
Association Technique pour e'Efficecité Energétique (1990) L'Eclairage Econome et Performant des Locaux Industriels et Tertiaires. Institut Francais De L' Energic, Paris
Authors V (1989) Ashrae handbook series 1989 fundamentals. SI Edition, USA
Boyle G (2004) Renewable energy: power for a sustainable future. Boyle Editor
Brunelli A (1987) Misure di Portata. Associazione Italiana Strumentisti, Milano
Capehart BL et al (2011) Energy management handbook, 8th edn. Fairmont Press, USA
Commission of the European Communities Directorate-General for Energy (1990) Energy in Europe energy for a new century: the European perspective. Office for Official Publications of the European Communities, Luxembourg
da Rosa AV (2009) Fundamental of renewable energy processes. Academic, USA
Decher R (1995) Energy conversion. Oxford University Press
Eastop TD, Croft DR (1990) Energy Efficiency for Engineers and Technologist. Longman Pub Group, Longman
Freeman H (ed) (1990) Hazardous waste minimization. McGraw-Hill Publishing Company, New York
Freris L, Infield D (2008) Renewable energy in power systems. Wiley, USA
Fuchs EF, Masoum MAS (2011) Power conversion of renewable energy. Springer, USA
Golusin M, Popov S, Dodic S (2013) Sustainable energy management. Academic, USA
Guiimot J-F, McGlue D, Valette P, Waeterloos C (1986) Energy 2000. Commission of the European Communities, Great Britain
Howell JR, Buckius RO (1987) Fundamentals of engineering thermodynamics. McGraw-Hill Book Company, Singapore
http://www.ec.europa.eu/clima/policies/f-gas
http://www.epa.gov/climatechange
http://www.gelighting.com
http://www.philipslumileds.com
Kalz DE (2011) Heating and cooling concepts employing environmental energy and thermo-active building systems, Fraunhofer Verlag
Kharchenko NV, Kharchenko VM (2013) Advanced energy systems. CRC, USA
Kreith F, West RE (1996) Handbook of energy efficiency. CRC, Boca Raton, FL
Loomis AW (1982) Compressed air and gas data. Ingersoll-Rand Company, Woodcliff Lake, NJ

G. Petrecca, *Energy Conversion and Management: Principles and Applications*, 315
DOI 10.1007/978-3-319-06560-1, © Springer International Publishing Switzerland 2014

Malloy T (1980) Thermal insulation design economics for pipes and equipment. McGraw-Hill Book Company, New York

Moran MJ, Shapiro HN (1988) Fundamentals of engineering thermodynamics. Wiley, Canada

Morvay Z, Gvozdenac DD (2008) Applied industrial energy and environmental management. Wiley-IEEE Press, USA

Petrecca G (1993) Industrial energy management: principles and applications. Kluwer Academic, Boston, MA

Ristinen R, Kraushaar J (2005) Energy and environment renewable energy sustainable. Wiley, USA

Robyns B, Davigny A, Francois B, Henneton A, Sprooten J (2012) Electricity production from renewable energies. ISTE and John Wiley & Sons, USA

Siuta-Olcha A, Cholewa T (2012) Energy saving and storage in residential buildings. Nova Science, USA

Smith CB (1981) Energy management principles. Pergamon Press, New York

Sorensen B (2007) Renewable energy conversion, transmission and storage. Elsevier, USA

Staff of Research and Education Association, Dr. M. Fogiel, Director (1989) The thermodynamics problem solver, REA, USA

The Institute of Electrical and Electronics Engineers—IEEE Power Engineering Society (1990) Publication Guide for Power Engineers, USA

Theodore Gresh M (1991) Compressor performance selection, operation, and testing of axial and centrifugal compressors. Butterworth-Heinemann Edition, USA

Thumann A, Metha DP (2008) Handbook of energy engineering, 6th edn. Fairmont Press, USA

Turner WC (1982) Energy management handbook. Wiley-Interscience, New York

Turner D et al (2012) Energy management handbook, 8th edn. The Fairmont Press, USA

Tyler G (1986) Hicks, power plant evaluation and design reference guide. McGraw-Hill Book Company, New York

Various Authors (1985) Manuale dell'ingegnere. Nuovo Colombo. Editore Ulrico Hoepli, Milano

Various Authors (1985) A user guide on process integration for the efficient use of energy. The Institution of Chemical Engineers, England

Various Authors (1987) The Essentials of Heat Transfer 1, Staff of Research and Education Association, Dr. M. Fogiel, Director, USA

Various Authors (1987) The Essentials of Heat Transfer 2, Staff of Research and Education Association, Dr. M. Fogiel, Director, USA

Various Authors, ENEA (1984) Metodologie di risparmio energetico. Editore Ulrico Hoepli, Milano

Westaway CR, Loomis AW (1977) Cameron hydraulic data. Ingersoll-Rand Company, Woodcliff Lake, NJ

William Payne F (1988) Integration of efficient design technologies. Fairmont Press, USA

Index